“John Mauchly and Presper Eckert we[illegible] of computing, but a variety of treacherous acts, bad breaks, and self-inflicted wounds kept them from getting the money and credit they deserved. Scott McCartney's interesting book restores this odd couple to their rightful place in technological history.”

—*San Francisco Chronicle & Examiner*

“Spare but compelling prose . . . The real contribution of *Eniac* is to document, in an engaging and informative manner, a significant benchmark in the historiography of computing.”

—*The Seattle Times*

“McCartney has performed an important service by rescuing this tale from obscurity.”

—*The Philadelphia Inquirer*

“Well researched and written.”

—*Library Journal*

“Despite the ubiquity of the computer, little is known about its origins. McCartney explores that irony . . . McCartney carefully researched documents, archives, and the personal papers of Mauchly and Eckert and interviewed surviving ENIAC participants. He also traces the concepts behind the computer through earlier efforts as far back as the 1600s, when Blaise Pascal developed an adding machine to track tax payments in France.”

—*Booklist*

“McCartney [chronicles] the technological challenges, the intellectual rivalries, the obsession and greed and, finally, in 1971, the fight for the patent for ENIAC, and with it the great accolade: ‘inventor’ of the computer.”

—*The Wall Street Journal*

continued on next page . . .

"A fascinating journey through the genealogy of the modern computer—detailing the science, business, governmental, and academic factors that have shaped the evolution of surely the most important development of the twentieth century. A compelling read for anyone and everyone interested in how today's computers came to be."

—Robert L. Crandall,
retired chairman and CEO,
AMR Corp/American Airlines

"A fluid history of the achievements and controversies surrounding the design and building of the world's first digital computer . . . written in a succinct style."

—*Kirkus Reviews*

"This lively account of the computer pioneers of another era not only fills a black hole in the history of technology, but demonstrates that the same chaotic and unpredictable creative processes that gave birth to the PC led, decades earlier, to the creation of the first mainframes. Without ENIAC, there could have been no Apple II, no IBM PC, no Macintosh."

—Walt Mossberg,
Personal Technology Columnist,
The Wall Street Journal

"McCartney vividly recreates the lives of John Mauchly and Presper Eckert, from their promising childhoods to their obscure deaths . . . [a] compelling tale."

—*The Dallas Morning News*

"Truly fascinating . . . lively . . . an amazing story of pure human endeavor."

—*Palo Alto Daily News*

Scott McCartney

ENIAC

The Triumphs and Tragedies of the World's First Computer

BERKLEY BOOKS, NEW YORK

A Berkley Book
Published by The Berkley Publishing Group
A division of Penguin Putnam Inc.
375 Hudson Street
New York, New York 10014

Text design by Ralph L. Fowler

PRINTING HISTORY
Walker and Company hardcover edition / June 1999
Berkley trade paperback edition / February 2001

The Penguin Putnam Inc. World Wide Web site address is
http://www.penguinputnam.com

Library of Congress Cataloging-in-Publication Data

McCartney, Scott.
ENIAC : the triumphs and tragedies of the world's first computer / Scott McCartney.
p. cm.
Include bibliographical references (p. 243–251) and index.
ISBN 0-425-17644-4
1. ENIAC (Computer) 2. Electronic digital computers—History. 3. Computer industry—United States—History. I. Title.

QA76.5 .M2945 2001
004.1—dc21 00-066788

PRINTED IN THE UNITED STATES OF AMERICA

10 9 8 7 6 5 4 3 2

For Abby and Jenny,

who will never know

a world without computers

Contents

ENIAC

INTRODUCTION

The Thinking Man's Game

In early 1997, Garry Kasparov, the Russian grand master of chess, squared off against "Deep Blue," an International Business Machines Corporation computer built with circuits designed specifically for the kind of computations—the kind of "thinking"—that goes on during chess. A year earlier, the chess champion had defeated Deep Blue, taking advantage of its lack of artistry. This time Kasparov, flustered by the craftiness of the upgraded machine and worn out by its speed, lost the six-game test of man versus machine.

Chess is a game not just of strategy but of struggle. It has captivated intellects for ages because it requires not just smarts but passion. It is an emotional battle, a creative challenge, a strategic war, and often a test of ego and fortitude. It is a human game, not a counting game. How could an inanimate box, an electric-powered number cruncher, possibly defeat the greatest human chess player?

Kasparov found himself overwhelmed by the computer's range. He changed his style of play in an attempt to outwit the mesh of solder and silicon, but the computer was always several steps ahead of him. Ultimately, the human let his emotions get the best of him. He missed opportunities and made mistakes. The computer, testing thousands of scenarios in seconds, was quicker and yet more careful. It was smarter, and it prevailed, winning the match in the final game.

What was most astonishing was how Kasparov described his opponent, having spent days matching wits with the box. The machine, he said, appeared to be developing a mind of its own, occasionally making moves that defied convention. "It showed a sign of intelligence," he reported after one long game. "I think that this machine understands that it's time to think."

Indeed, computers have progressed to the point where they can be taught to "understand" the unspoken machinations of the chess match, where they can "think" faster and better than humans can. To many, Kasparov's loss was a seminal event, one that marked the end of man's intellectual superiority over machines.

In reality, the event was simply the culmination of more than fifty years of computer science achievement, an era of blistering-fast miniaturization and exponential advancements in speed that together have yielded incredibly cheap, powerful computers. They have changed the world in profound and innumerable ways, controlling business, travel, government, communications, and even our ovens and lawn sprinklers. Economists reason that computers have allowed the enormous economic expansion in the United States since the 1960s by delivering productivity gains across society. Without computers, it would have been difficult for the nation to keep up with postwar growth, unable to muster enough manpower to sustain a prolonged expansion. But the machines arrived at exactly the right moment to prevent the economy from being buried in paper and restrained by the limitations of the human brain. Computers have produced a higher standard of living for the United States and the rest of the industrialized world. Indeed, cheap, powerful computers are spurring rapid development in Asia, Latin America, and Africa. Now these tools, these machines created by man, have actually gone much farther. They possess intelligence; they can outthink the grand master of chess.

How did this dazzling brainchild begin? Who first thought of the notion of using electrons to compute? How preposterous a notion it must have been, to run an electric current through a machine and expect it to calculate, analyze, and ponder. Electricity lights things; it moves things. It's energy, not intelligence; waves, not thoughts. Machines don't have neurons; the human brain can't be mechanically duplicated. How, then, could electricity be used to "think"?

The invention of the computer ranks as one of the

greatest achievements of the century, indeed, the millennium. Yet the inventors remain obscure, and the story of how their invention came to be has been largely overlooked. Most people know who discovered electricity, who created the lightbulb, invented the telephone, flew the first airplane. The titans of the computer industry have become household names: Thomas Watson Sr., Bill Gates, Steve Jobs, Michael Dell. But none of them invented the first computer. Even though computers have made memory a thing you can buy at stores, we have already forgotten the origins of these remarkable machines.

Washington Post columnist Bob Levey tells a story of an Alexandria, Virginia, mother who quizzed her nine-year-old son on inventors for a school project. The boy tagged Thomas Edison for the lightbulb and Eli Whitney for the cotton gin. "Who invented the computer?" she asked, pushing beyond the school's list. The boy thought for a minute, then replied: "Radio Shack."

It's not a bad answer. Yet three decades before Radio Shack began selling some of the first personal computers, the computer was born as a giant behemoth, labeled in the early press accounts as the "Giant Brain." It consumed enough electricity to power a small village, and it performed what we now consider sophomoric calculations at Model T–like speeds. The importance of the first machine is often minimized when computer experts make astonishing comparisons—for example, how the computing muscle of the apartment-size first computer would now fit on a thimble of circuitry. We race ahead to the next new model that is faster and smaller and more powerful, discarding computers that are but eighteen months old as retired relics. And we forget the consequence that the first computer had on today's wonder boxes.

To many people, the father of the computer is a scientist named John von Neumann, already worshiped in academic circles for his contributions to mathematics and logic before he got involved in the development of Giant Brains. The basic architecture of the computer became known as the "von Neumann architecture," and that term—just like the design it describes—remains in use today.

John von Neumann didn't invent the computer, however. The distinction rightly belongs to two men at the University of Pennsylvania, Presper Eckert and John Mauchly. They built ENIAC, the first digital, general-purpose, electronic computer—the first Giant Brain. ENIAC was a bus-size mousetrap of forty nine-foot-tall cabinets filled with nearly 18,000 vacuum tubes and miles of wiring. It was developed as a weapon of war, a machine that could calculate trajectories for World War II artillery guns. Though unwieldy, ENIAC was wired with enough innovation and genius to provide the spark for computer development. The genealogy of the modern computer begins here, with ENIAC, Eckert, and Mauchly.

Yet because ENIAC operated under army secrecy all

of its life, it was little known except in computing circles. What's more, ENIAC was buried in the annals of history under controversy, jealousy, and lawsuits. Amazingly, Eckert and Mauchly, who not only built the first computer but also founded the world's first computer company, secured neither fame nor fortune. They have been largely forgotten, as if deleted from the hard disk of computer history.

This book is the story of ENIAC, Eckert, and Mauchly. The first electronic digital computer was birthed by people toiling under tremendous wartime pressure, a group of believers and skeptics who took a chance on a young professor's "absurd" idea because lives were at stake. The team that built the first computer ate, slept, and lived with the machine, devoting their lives to a project that, according to the experts of the day, had little chance of success. They faced technical hurdles like how to wire rings of vacuum tubes to "count" numbers without making errors or simply burning out. How would the circuit know when to stop? How would it transmit the answer? They faced logical issues like how to get those rings to carry digits if a sum exceeded 10, and how to wire up the process of taking a square root. How, they wondered, could you "program" the machine? Many said what they were attempting to do was technically impossible; few understood what the new machine might be capable of doing.

Yet Eckert and Mauchly persevered, and prevailed with ENIAC. Sometimes they succeeded because they were brilliant in their design and meticulous in their execution.

Sometimes they went with their hunches and hoped for the best. They tore up electrical schematics when last-minute glitches were found. When engineering roadblocks rose before them, they improvised. When materials were in short supply, they scrounged and scavenged. And when a fire almost destroyed the project, they redoubled their effort. When ENIAC ran its first calculation, even Eckert and Mauchly were stunned.

Their marathon had been paced by the excitement of creating not just a new machine but a new science. They focused not just on the urgency of war but on the larger goal of creating something that would last, something that would change the world. Amazingly, they understood the impact the computer would have and quite accurately predicted its use in business and government. They even opined that small computers would one day fit on a desktop and be far cheaper and far more powerful than their invention.

Once World War II was over, however, the camaraderie of a wartime mission splintered into competition and pettiness, and some of the computer's pioneers turned on one another, muddying the whole beauty of its creation. Mauchly and Eckert's success story has a tragic ending. For all their creative genius, the builders of the first computer turned out to be lousy marketers and poor businessmen. They had vision but lacked the acumen to profit from their technological accomplishments.

As a new century begins to define its own course, with

the computer as a central player in our lives, this story needs to be told now more than ever. It is based on a close review of documents, research from the archives of several institutions, interviews with surviving participants, and a study of resources never before examined, including the personal papers of both John Mauchly and Presper Eckert, boxes of which remain in attics at family homes. *ENIAC: The Triumphs and Tragedies of the World's First Computer* sheds new light on what actually happened and challenges some of the oldest perceptions about the development of the computer, bringing clarity to this seminal event in the worlds of technology and business. In the end, like Deep Blue's victory over Garry Kasparov, this is a tale that says more about man than machines.

CHAPTER ONE

The Ancestors

"You've got mail!" The phrase has become part of our everyday lexicon, a new signpost of popular culture, one that defines an era, much like "I am not a crook" or "I can't believe I ate the whole thing."

Today, a wired America, physically and emotionally, communicates on-line. We have taken Morse code to a new extreme. We can send the equivalent of a library full of information around the globe in bits and bytes, a form of dots and dashes, as fast as energy can move through wires. None of this would be possible, of course, without computers, which provide the backbone of the Internet, the telephone systems, and just about every other mode of communication, transportation, commerce, and government. The computer today has become the heart and lungs of our society, powering and sustaining our lives efficiently and safely . . . until it crashes.

We rely on the computer to process, to collect, to store, to track, to decide, and to resolve. It is everywhere. What does the grocery store do when the computer goes down? Could we communicate in the office without E-mail? The computer is controlling our lives, if not driving them. And to think, the computer came into existence just over fifty years ago. Generation after generatiaon of computer has been born and, within a few years, died unceremoniously, but this rapid growth has occurred without much careful planning.

So shortsighted was computer development that nobody planned for the millennium, and businesses and governments are spending billions of dollars to ready computers for the next century. Herbert Kelleher, the chairman and chief executive of Southwest Airlines, recalled the first meeting his company had on the Year 2000 problem, when computer engineers warned that the airline's computers might think it was January 1, 1900, instead of January 1, 2000.

"So what?" Kelleher asked. "Won't *we* know?"

Turns out, it's more important that the computer know. To save precious electronic real estate, shortsighted software developers abbreviated years to two digits instead of using all four. When the year rolls over to 2000, some computers will become confused—and confused computers are likely to shut down. Already, credit cards with expiration dates past 1999 have been unusable in some stores because computers rejected them as "expired." For Southwest and other airlines, the Year 2000 problem may manifest itself in many ways. For example, Federal Aviation Administration computers keep airplanes from crashing into one another. But if the computers themselves crash and airport control towers are unable to use them, it won't be safe for the planes to fly.

The Year 2000 problem looms so large because society depends on computers to make sense of information, to turn data into intelligence.

The power of the computer is that it brings order to chaos. We know that even in nature's randomness, there often is a pattern—we just don't see it. Computers help us sort and shape the world in brand-new ways. They get their intelligence from their power to count, sort, organize, and compare. By brute force—speed and capacity—computers enable government to provide Social Security checks to millions, manufacturers to efficiently run factories, stock exchanges to process billions of transactions, schoolchildren to study frogs in Australia. In each case, the computer is able to order the jumble of data, to make intelligence out of information.

In the history of civilization, truly important advances have come when one is able to bring new order to the world. Around 1600, the Italian astronomer, mathematician, and physicist Galileo mathematicized the physical sciences, developing formulas to explain how the physical world worked. That was a watershed event in scientific history, opening new doors to knowledge. Before Galileo, scientists investigated nature and made measurements, but not much more. With his telescopes, Galileo gained a new understanding of the solar system. His conclusions were so radical he was even put on trial by the Inquisition for saying the sun was the center of the universe and the earth revolved around the sun. Galileo timed the oscillations of a swinging lamp, discovering the isochronism of the pendulum, and that the path of a projectile is a parabola. His findings foreshadowed Sir Isaac Newton's laws of motion. But what was most important, Galileo showed how to create order out of chaos.

The development of the computer is intertwined with this quest for understanding, this search for order. The machine was designed to sort and organize information, and solve problems. Some investors simply wanted a tool to handle drudgery and repetitive work; others had grander visions of mechanizing the mind. Often the computer is thought of as a twentieth-century phenomenon, but its genealogy is sprinkled within three hundred years of scientific advancement.

Pascal to Babbage: Getting into Gear

In 1642, the year Galileo died, a young man named Blaise Pascal developed an adding machine for his father, Etienne, who was a high official in France, where a way of reorganizing taxation was needed in the face of revolt. The government wasn't keeping up. Pascal fashioned an eight-digit mechanical calculator called the Pascaline that could perform addition when the operator turned a series of rotating gears. The brass rectangular box had eight movable dials; as one dial moved ten notches, or one complete revolution, it moved the next dial one place. When dials were turned in proper sequence, a series of numbers was entered and a cumulative sum obtained. But because the gears were driven by a complicated set of hanging weights, the machine could work in only one direction; hence it could handle addition but not subtraction. By 1652, fifty prototypes had been produced, but few machines were sold. Nevertheless, Pascal demonstrated the importance of calculating machines, and he proved two axioms of computing history. First, the young will lead—Pascal was only nineteen at the time of his invention, making him perhaps the first teenage whiz kid in computer history. Second, new technologies will catch on slowly unless there is a clear and urgent use awaiting them.

Gottfried Wilhelm Leibniz, a German mathematician and philosopher, took Pascal's wheel one giant turn farther, and learned the same lesson. Born in 1646, four years after Pascal's invention, Leibniz produced a machine capable of multiplication, division, and subtraction as well as addition. Leibniz wanted to free people from the slavery of dull, simple tasks. Partly by studying Pascal's original notes and drawings, he was able to refine the Pascaline into a calculator so ingenious that the design would last for centuries in mechanical desk calculators. Instead of a simple, flat gear, Leibniz used a drum for his wheel, putting on the drum nine teeth of increasing length to make steps. He called his machine the Stepped Reckoner. The user turned a crank once for each unit in each digit in the multiplier, and a fluted drum translated those turns into a series of additions. In 1673, the Leibniz machine was completed and exhibited in London. But it was so far ahead of its time, it was not appreciated until after Leibniz's death. Leibniz's calculation machine couldn't be a success because economics were against it, Vannevar Bush, one of the foremost American scientists, wrote in 1945. There were no savings in labor: Pencil and paper were still cheaper and faster.

Not until 1820 or so did mechanical calculators gain widespread use. By then, the industrial revolution was under way, mass-production manufacturing was in hand, and the public had an enormous hunger for anything mechanical. The steam engine was the prime technology of the time, iron suspension bridges were being constructed, and in England, the Stockton and Darlington Railway was inaugurated in 1825. At the same time, Charles Babbage, a founding member of the Royal Astronomical Society and part of England's intellectual upper crust, conceived of the computer, though he didn't call it that and his device was far from an actual computer. Babbage, who was a mathematician, astronomer, and economist, recognized that an increasingly industrialized society needed better, more accurate methods of calculating. One particular problem was that tables used in maritime navigation, astronomy, and industry were often filled with inaccuracies, and doing calculations yourself required weeks and months of mindless number crunching. Babbage knew that many long computations consisted of operations that were often repeated, so he reasoned he could create a machine that would do those operations automatically. In effect, he wanted to mechanize mathematics, just as machines were converting factories to mass production.

Following Pascal and Leibniz, he designed a machine with a set of adding mechanisms—gears and wheels—that could calculate logarithms and even print the result by punching digits into a soft metal plate for a printing press. He called his room-size contraption the Difference Engine and even designed it as a steam-powered device. Its logic was conceptually simple; the machine would produce mathematical tables by comparing differences in numbers. But mechanically, it was complex.

Work began on the full machine in 1823, and it took ten years to produce a working prototype. Funded in part by the British government, the project was plagued by cost overruns and production difficulties. Unfortunately, the plans for the Difference Engine exceeded the manufacturing capabilities of the nineteenth century. (The task was later likened to a pharaoh having detailed designs for an automobile and trying to build one using the tools, materials, and expertise of the day.) As it turned out, the Difference Engine was never built—partly because Babbage had already fallen into a common trap in computer development. He had come up with a better idea before he completed his first machine, and he turned his attention away from the project already in progress.

In 1833, Babbage unveiled his idea for a new machine, the Analytical Engine. Remarkably similar in design to modern computers, the device could perform any kind of calculation, not just generate tables. Babbage had struggled to make the Difference Engine an automatic device that could take the result of one calculation—one point on the table—and use it to begin the next calculation, for the next point on the table, a process he called "eating its own tail." He found his answer in one of the most brilliant industrial developments of the day, the Jacquard loom, which became the basis for the Analytical Engine.

Joseph-Marie Jacquard, a Frenchman, had devised a loom that would weave by following a pattern laid out in a series of cards with holes punched in them. Intricate patterns of flowers and leaves could be repeated or revised, all by changing the perforated cards. Babbage decided to make a calculating device driven by two sets of cards, one to direct the operation to be performed, and the other for the variables of the problem. His assistant, Augusta Ada King, countess of Lovelace, who was the daughter of the English poet Lord Byron, said the Analytical Engine "weaves algebraical patterns." By placing cards in an orderly series, Babbage could organize various computational tasks—he could *program* his computer.

The Analytical Engine had another very important concept built into it: It could hold partial results for further use. In effect, it could *store* information. The machine had a "mill," a term Babbage borrowed from textiles, which consisted of two main accumulators and some auxiliary ones for specific purposes. It functioned as the central processing unit—numbers could be added together in the accumulators, for example. In the textile industry, yarns were brought from the store to the mill, where they were woven. So it was with the Analytical Engine; numbers were brought from storage and woven in the mill into a new product.

Most important, the Analytical Engine had a key function that distinguishes computers from calculators: the conditional statement. The Analytical Engine, Babbage said, could reach a conditional point in its calculations and see what the current value was. If it met one condition, say, greater than 0, the Analytical Engine would proceed one way. If the interim value was less than 0, the Analytical Engine could take a different course.

The Analytical Engine was a computing device so brilliant Babbage was derided as a lunatic. The British government refused to fund such a radical contraption, although the government was undoubtedly perturbed that the Difference Engine it had helped fund a decade earlier was never completed and was now defunct. Babbage was largely ignored by contemporaries after he made the proposal for the Analytical Engine. There were many impediments to Babbage's machine. Not only was there no real need for it, but it would not have been very fast. Nor was the technology available to build all the gears and wheels to exacting tolerances. But Babbage's idea would eventually become a reality.

The Dawn of "Data Processing"

The need to handle numbers in an increasingly industrial society did ultimately spur further development. "Data processing" became a mechanical function in 1890 when Herman Hollerith of the Massachusetts Institute of Technology built a punch-card tabulator for the U.S. Census Bureau. The census of 1880 had taken nearly seven years to count, and the Census Bureau feared that with an expanding population, the 1890 census might take ten years, rendering it useless. Hollerith's proposal wasn't inspired by Jacquard's loom or Babbage's Analytical Engine. Instead, the idea came to him as he watched a train conductor punch tickets so that each ticket described the ticket holder: light hair, dark eyes, large nose, and so forth. The ticket was called a punch photograph, and Hollerith recalled to biog-

letter that his inspiration was simply to make a punch photo-rapher Geoffrey D. Austrian that his inspiration was simply to make a punch photograph of each person for the census. Herman Hollerith was far more practical a genius than his dreamy data-processing predecessors.

Hollerith didn't use punch cards to instruct his machine the way Babbage had envisioned. Instead, Hollerith's cards stored the information. A punch at a certain location on the card could represent a number, and a combination of two holes could represent one letter. As many as eighty variables could be stored on one card. Then the cards were fed through a tabulating and sorting machine. The tabulator had a plate of 288 retractable, spring-loaded pins. A pin encountering a hole completed an electrical circuit as it dipped through, and the electricity drove a counter, not unlike the gears and wheels in Pascal's original calculator. The circuit could also open the lid of one of twenty-four sorting compartments and send the card to the next phase of operation. If the bureau wanted to count men with three children, the cards could be sorted accordingly. For the first time, electricity and computing were converging as intelligence inside a machine. It made for a smarter machine because it made for a faster machine.

Initial results from the 1890 census were tabulated in just six weeks, but Hollerith's punch cards were hardly an instant sensation. The Census Bureau declared that the nation's population totaled 62,622,250. Pundits had been hoping that the country had grown to 75 million people, and in the initial shock and disappointment, the machine was blamed. "Useless Machines," the *Boston Herald* declared. "Slip Shod Work Has Spoiled the Census," the *New York Herald* chimed in.

Hollerith, of course, had the last laugh. The entire census was completed in less than three years, at an estimated savings of $5 million. It was also far more accurate than previous tallies. Hollerith had figured out a way for his card-tabulating machine to perform multiplication, enhancing its capabilities. Even at the racetrack, Hollerith's ideas took the form of mechanical calculators called totalizators—or tote boards—that continuously added bets and figured odds for each entry. "Real-time computing," or at least calculating, was born. His invention was the backbone of a new company he formed in 1896, Tabulating Machine Company, which through acquisitions ultimately became International Business Machines Corporation. Punch-card machines became the gold standard for data processing. The technology opened new avenues for government. It enabled the Social Security Administration to come into existence in 1935, using punch cards to store wage data on every working American.

Already, there was pressure from scientists to produce faster machines, which would allow for more complicated calculations and more accurate tasks. The 1930s was a time of rapid advancement in "calculating machines"—mechanical contraptions that began making life faster and easier—and MIT was at the center of the leading research.

In 1930, MIT's Vannevar Bush made an enormous breakthrough by developing a machine called the Differential Analyzer. Bush's machine, which used a mathematical method different from that of Babbage's Difference Engine, was full of shafts, wires, wheels, pulleys, and 1,000 gears that appeared to be the work of overzealous elves rather than an effulgent scientist. Everything was driven by small electric motors that had to be carefully adjusted for accuracy of calculations. The Differential Analyzer had to run at slow speeds, because disks could slip and throw numbers off. It measured movements and distances to yield the results of computations based on those measurements.

The machine had input tables, into which operators fed numbers and functions. Its gears could be adjusted to introduce constant functions: Want to double a result somewhere along the calculation? Simply adjust the gears. There was a unit to do multiplication and an output table that could display results in graphs. Different units could be connected in a certain sequence by long shafts, and how the connections were made programmed how the machine would solve the problem. It looked like a huge erector set, wider than the length of a banquet-size dining table. Programming it was a major chore, since all the interconnections had to be undone and redone.

For all that, the Differential Analyzer could solve only one type of problem—differential equations. Differential equations are an important type of calculation involving rates of change and curves that are used to describe happenings in the physical environment, such as the area lying under a curve or the path of a projectile. The formula

$$x^2\left(\frac{d^2y}{dx}\right) + x\left(\frac{dy}{dx}\right) + y(x^2 - n^2) = 0$$

is an example of a differential equation used to describe vibration and periodic motion. Once assembled on the Differential Analyzer, substituting any number for *n* would yield the values of the other variables in the equation. What's the position of the projectile after two seconds? After nine seconds?

The Differential Analyzer was "analog" rather than "digital." That is, it worked in continuous waves rather than specific digits. Like a clock with hands, an analog-computing device gets to the right answer by moving a distance specified by the numbers in the problem. Digital devices, on the other hand, use only discrete numbers, such as those on a digital clock. A speedometer is analog; an odometer is digital. Analog devices are less accurate because the readout is an estimate. In addition, gears wear down, wheels slip, and the machines, over time, grow more inaccurate.

Despite its inherent inaccuracy, the Differential Analyzer was still far superior to other calculating devices, and it became the most sophisticated computational tool for scientific research. It helped lay the groundwork for the notion that if you wanted something fast, you had to make it electric.

Still, electricity was only the muscle, not the brains. As several people around the world were pressing on with more sophisticated calculating machines in the 1930s, a key technological innovation came from an unlikely source: the telephone company.

Electric Calculators Make Their Mark

Bell Telephone Laboratories had been working for years on the science of turning information—the numbers dialed into a telephone, the sound to be transmitted—into electrical signals. To route each call, the telephone company has a network of switches. Early switches required operators to physically connect lines with patch cords. Later, devices called relays were developed to make the connections much faster. Relays are electromechanical gadgets with an armature inside that opens or closes based on electrical current. Like a light switch, they can actually represent information depending on their position. Thus, making a computing machine with switches was a natural pursuit for Bell Labs.

In the mid-1930s, one of Bell Labs' research mathematicians, George R. Stibitz, began using telephone relays to register either a 1 or 0, depending on the position of the relay. He created a circuit of relays, called a flip-flop, which could count based on the flow of electricity. The circuit had lamps that glowed for the digit 1 and were dark for the digit 0. Stibitz's "breadboard" circuit, first assembled on his kitchen table, could count in binary arithmetic, where our traditional base-10 numbers are coded as a series of 1s or 0s. In base 2, four relays could represent ten digits: 0 was 0000, 1 was 0001, 2 was 0010, 3 was 0011, 4 was 0100, 5 was 0101, 6 was 0110, 7 was 0111, 8 was 1000, and 9 was 1001.

Stibitz had realized that the telephone relay—the on-off switch—was well suited to the binary arithmetic he had learned in algebra class. If you could sling enough relays together, you could make a very fast calculator. Instead of numbers being represented by slow-moving wheels and drums, they could be represented by fast-moving electricity.

In the fall of 1937, Stibitz drew up plans for a machine to multiply complex numbers and handle computations the company's engineers needed to better construct voice circuitry. The device, creatively called the Complex Number Computer, couldn't be programmed; once numbers were fed in, it multiplied them automatically. It couldn't do further calculation with the result on its own. Construction began at the old Bell Laboratories Building on West Street in Manhattan in April 1939 and was completed in just six months. The machine was an improvement over mechanical desk calculators of the day, and it operated for nine years. But its real value lay not in what it could do but in what it showed could be done. The flip-flop would prove to be a key invention in the development of the computer. It was the means by which electricity could be converted into numbers, and circuits could ultimately be made to process intelligence.

The same year Stibitz was drawing up his plans, Howard Aiken at Harvard was working on computing machines when he stumbled upon Babbage's work. He began working on machines with IBM that would become electromagnetic versions of the Hollerith punch cards. It seemed like a strong alliance. IBM, funding Aiken's work, had itself grown out of Hollerith's Tabulating Machine Company. Hollerith had sold his firm to businessman Charles Ranlegh Flint, who merged it with a maker of shopkeeper's scales and a manufacturer of workplace time clocks to create Computing-Tabulating-Recording Company, or CTR. In 1924, CTR's president, Thomas J. Watson Sr., renamed the conglomerate International Business Machines and issued his famous proclamation: "Everywhere . . . there will be IBM machines in use. The sun never sets on IBM."

It was that hunger for ubiquity that drove IBM to fund not only its own development laboratories but also university research at several institutions. IBM, faced with a rapidly changing industry and growing competition, spent liberally so it could stay on top of developments that might lead to new office machines. One of the premier projects was work at Harvard on automatic computing machines—especially Howard Aiken's Mark I.

The Mark I was a slow electromechanical giant that

borrowed some of Babbage's principles and used wheels with ten positions on them to represent digits. In fact, Aiken had come across a fragment of Babbage's calculating engine in the attic of Harvard's physics lab, a donation in 1886 by Babbage's son, Henry. Aiken became fascinated with Babbage and began to see himself as the new carrier of Babbage's baton.

The Mark I, constructed by IBM, was a five-ton calculator. The machine's components were lined up and driven by a fifty-foot shaft. Altogether, it had some 750,000 moving parts, and when it churned, observers said it sounded like the roar of a textile mill. Those suffering the din probably would not have appreciated just how fitting that comparison actually was. The beauty of the Mark I was that it was the first fully automatic machine: It could run for hours or even days without operator intervention. Programs were entered on paper tape, and if the same functions were required over and over again, the tape was simply glued together. But Aiken, or IBM, apparently hadn't studied Babbage closely enough; the Mark I had no "conditional branch"—the If . . . then statement of computing—even though Babbage designed such a function into the Analytical Engine. And the machine had one huge drawback: Although it could undertake calculations never before mechanized, it was painfully slow.

Now part of the war effort, Mark I undertook computations for the U.S. Navy. As its public unveiling neared in

1944, Aiken began clashing with IBM. Thomas Watson, now the company's chairman, wanted to make a splash with his investment. So he commissioned a designer to spruce up the look of the machine, and it was fitted, over the objections of Aiken, who wanted easy access to all components, with a shiny cabinet of stainless steel and glass with the IBM livery. Then, at the dedication ceremony at Harvard in 1944, according to computer historians Martin Campbell-Kelly and William Aspray, Aiken took credit as the sole inventor of the calculator. He acknowledged neither IBM's underwriting nor the contribution of the firm's engineers, who had taken Aiken's rough concepts and hammered them into a functioning machine. Watson was said to have been shaken with anger and became determined that IBM would undertake its computing research without Harvard.

Aiken himself declared that Mark I was "Babbage's Dream Come True." While Mark I was indeed the kind of machine Babbage had proposed one hundred years earlier, it was, according to Campbell-Kelly and Aspray, perhaps only ten times as fast as the Analytical Engine would have been. The Mark I was so slow it was a technical dead end, soon to be eclipsed by electronic machines thousands of times faster than Babbage's dream. Aiken had used electricity to drive his counting and calculating components, but he had not made the leap to using electricity as the counting and calculating component itself. That leap—using electricity to think—would come in a most unusual way.

CHAPTER TWO

A Kid and a Dreamer

While still a youngster in knee pants in 1919, John William Mauchly had a habit of reading books and magazines in bed late into the night. The Mauchly family lived in a modest, four-bedroom, one-bath frame house in Chevy Chase, Maryland, long before it became a swank power enclave. The staircase had a landing three steps above the ground floor, and from one or two stairs above the landing, John's parents, Sebastian and Rachel, could see under his bedroom door. If the light was on past bedtime, trouble was at hand.

Young John had a solution to his predicament. He placed a sensor under a loose board on the staircase landing, then wired it to a small light in his room. Whenever someone stepped on the landing, the light went out in his bedroom, giving him time to turn off the main light before his mother or father could see under the door. When an inquisitive parent went back across the landing, the light toggled again, giving the all-clear signal.

By the time he had pulled that stunt, John Mauchly's wiring career had been under way for a while. At age five, he had taken a dry cell battery, a lightbulb, and a socket and fashioned a flashlight to explore a dark attic at a friend's house. The friend's mother was so afraid the new creation would start a fire, she gave him a candle to use instead. In grade school, Mauchly earned money installing electric bells in place of mechanical bells. As crews dug waterlines in the new suburb, Mauchly laid intercom wires in the trenches so he could communicate with friends. When neighbors had trouble with their wiring—it was only forty years after Edison invented the incandescent lightbulb, and electricity was not yet a mature public utility—they called John Mauchly. On the Fourth of July, friends gathered as Mauchly set off fireworks and even a pipe bomb by remote control. One April Fool's Day, Mauchly wired the front doorbell so that callers received a mild shock when they rang the bell. He built alarms in his basement workshop and even a model elevator with a lever to make it move up and down.

The time around 1919 was an age of discovery, not just for young John Mauchly, who was born on the cusp of the new century, but for the whole world. The secrets of the atom, the basic unit of matter, were rapidly being uncovered. Albert Einstein had completed his theory of relativity. The Great War was over, and the United States was heading into the Roaring Twenties. A few years earlier, Sebastian Mauchly, whose grandparents had emigrated to Ohio from Zurich, Switzerland, in 1839, had earned a Ph.D. in physics, the hottest of academic fields. He left his job in Cincinnati as a high school principal and in 1916 moved his family to Chevy Chase, where he became chief physicist at the Carnegie Institution's Department of Terrestrial Magnetism, researching what makes the earth magnetic and how lightning works. Chevy Chase was a community of scientists, with officials from the National Bureau of Standards, the U.S. Weather Bureau, and other government agencies. And in that environment, John Mauchly—a restless child with a sign over his bed that said, "What should I be doing now?"—thrived. He earned near-perfect scores in high school, was a whiz at math and physics, and was editor of the school paper his senior year (1925).

Realizing that scientists—even physicists—were generally poorly paid, Sebastian encouraged his son to become an engineer. Engineering was not as challenging a pursuit, perhaps, but it paid better. John won an engineering scholarship to Johns Hopkins University in Baltimore, Maryland, but by his sophomore year, he was bored. Engineering, he recalled in interviews later, was "cookbook" stuff. You designed a beam for a certain load, then looked in the U.S. Steel Company book for how much steel you needed, and how many rivets. "Physicists—those were the boys who were going to have fun," Mauchly said.

John Mauchly (Annenberg Rare Book & Manuscript Library, University of Pennsylvania Library)

After two years of undergraduate engineering, he switched directly to the Hopkins graduate physics program and studied molecular spectroscopy, spending hours upon hours calculating the molecular energy of gases. To crunch his numbers, Mauchly used the Physics Department's Marchant desk calculator, a souped-up mechanical adding machine that could perform large-scale multiplication or division, plus addition and subtraction, by pressing the desired buttons and then pulling a big handle. He graduated with a Ph.D. in 1932.

By this time, John Mauchly was a married man. In 1930, he had wed Mary Augusta Walzl, a mathematician, and the couple would eventually have two children. But while his personal life was satisfying, he had trouble landing his first job. The nation was deep in the Great Depression, and the services of newly minted Ph.D. physicists, as his father had feared, weren't needed. Mauchly stayed on at Hopkins for a year as a research assistant, then finally was offered a job at Ursinus College in Collegeville, Pennsylvania. The position paid $2,400 a year, but the faculty had already agreed to give back 10 percent of their salaries to keep the liberal arts school going.

Mauchly was a dashing, if often rumpled, professor with brown hair and hazel eyes. Standing just a half inch under six feet and weighing 180 pounds, the long-limbed Mauchly was well read, soft-spoken, and whimsical. An inveterate teacher who even later in life would offer long lectures, and pause in conversations to explain theory and laws, Mauchly could be easily stereotyped as mild mannered and forgetful, though he was a ferocious record keeper.

Ursinus was a school training the odd combination of premedical students, who needed one course in physics, and high school teachers, who likewise needed only the basics. Yet with expectations low, Dr. Mauchly quickly distinguished himself as an extraordinary teacher who could make science come alive. He became so well known for his "Christmas lecture"—the last lecture before midyear exams—that professors from other universities traveled to Ursinus just to see what all the fuss was about, and the lecture had to be moved to the biggest hall on campus. One time, Mauchly put a skateboard atop a lecture desk and demonstrated Newton's three laws of motion. If he moved to the right, for example, the board moved to the left, and if he moved to the left, the board moved to the right because for every action, there is an equal and opposite reaction. Another time, Mauchly wrapped Christmas presents in colored cellophane and then demonstrated spectroscopic principles by showing the audience how to find out what's in a package without opening it. Another year, he dressed as a Russian on roller skates and spun around, putting his arms out to slow down and pulling them in to speed up.

On the side, Mauchly pursued his favorite obsession—predicting the weather. Using atmospheric data he had access to through old Chevy Chase connections at the Weather Bureau, he tried to see if weather patterns could be predicted mathematically. The theory, proposed by Lewis F. Richardson in England and others, was that the atmosphere followed specific rules, and if those rules could be identified, weather could be accurately predicted. The pursuit led Mauchly to bone up on statistics, and soon he was publishing articles and making presentations to academic societies on the application of statistical techniques to questions like, Does the sun influence our weather?

Mauchly bought a desk calculator similar to the one he had used at Hopkins for seventy-five dollars from a bank that had failed. With dozens of Ursinus students adding up data on weather maps from across the country and Mauchly running calculations on rainfall reports, he hypothesized there was evidence that rainfall in the United States had a kind of periodicity to it that was related to the rotation of the sun. But with so many variables and so much data, it took days and months and even years to perform the necessary conclusive calculations. At that point, he realized that if he was ever going to unlock nature's meteorological puzzles, he needed a better, faster calculating machine.

"There Must Be Something Better"

His tenacity with the problem of weather prediction was typical of John Mauchly, who would not let go of a topic until he had taken it as far as it could go. "I have a sort of stubborn streak in me," he confessed late in life in a videotaped interview with his friend Esther Carr. "Some people call it champion of the underdog."

That stubborn streak also kept him playing with wiring, circuitry, and calculating machines. The tools available were slow and limited. "I kept thinking there must be something better," he said. Mauchly built all kinds of gadgets with his Ursinus students, including a device called the Harmonic Analyzer that measured changes in data, something useful to his weather-prediction problem. He took students on field trips to see actual atomic research, and learned at Swarthmore College about work with vacuum tubes, which were used to distinguish between electrical pulses as close as one-millionth of a second. Physicists at

Swarthmore were using vacuum tubes to count cosmic rays at incredible speeds.

Vacuum tubes had been the key to the early days of the electronic age. The first tube was constructed in 1904—a variation on Edison's lightbulb of 1883. Vacuum tubes allowed engineers to easily regulate or amplify the flow of electricity, enabling devices such as radios, televisions, and telephones to work efficiently. The simplest ones had only two elements inside the vacuum: an emitter and a collector. With those elements, you could control electricity at very rapid speeds by using the tube as a kind of on-off switch. The presence of electricity turned the tube on; the absence kept it off. The tube, then, became the light switch of the circuit—and it was much faster than any mechanical switch or toggle. That ability to turn off and on very, very rapidly led Mauchly to consider a new kind of counting. What if you could generate your own pulses to represent numbers, like the pins on the desk machine, and control them and count them with vacuum tubes just as in the physics experiments? Then you would have a fast way to calculate numbers—to compute. That would clearly be much faster than the mechanical methods available at the time. A calculation could be performed in two-hundredths of a second, Mauchly estimated, meaning complex problems could be solved quickly and efficiently.

In 1939, Mauchly took a night course in electronics to begin learning more about circuits, and he went to the New York World's Fair, where he saw an IBM electric cryptographic machine that used vacuum-tube circuits for coded messages. He began ordering materials from around the country, writing to Supreme Instruments Corporation, Greenwood, Mississippi, in September 1939 asking about switches because "I am intending to construct an electrical calculating machine." As his tinkering continued, he learned from a local hardware store of a new type of fuse with a little red indicator light—a neon bulb—that would illuminate when the fuse was burned out. Those miniature General Electric bulbs proved far cheaper than vacuum tubes, and Mauchly began ordering them in bulk at eight cents apiece. They were a thousand times slower than vacuum tubes but could distinguish between pulses one one-thousandth of a second apart—plenty fast for "garage" experiments.

By now, Mauchly had picked up the trail blazed by calculating machine pioneers, though he was so busy wiring his circuits and crunching his weather data that he wasn't aware of most of the theory and principles articulated long before. It didn't occur to him to study calculating machines of the past. To Mauchly, this was a new field, lying in front of him. Like a racehorse, he had his blinders on, and he was galloping down the track.

Soaking up all he could from his contemporaries, Mauchly drove to Dartmouth College in New Hampshire in 1940 to a meeting of the American Mathematical Society, where Stibitz demonstrated his electromechanical calculator, the Complex Number Computer. Stibitz explained how relays used in telephone circuits were better than gears and wheels in calculating machines; Mauchly already knew vacuum tubes were better than relays because he had seen tubes working in physics experiments at Swarthmore. After the demonstration Mauchly chatted with Norbert Weiner, a prominent MIT mathematician, and the two agreed that electronic computers were "the way to go." After that, Mauchly stepped up his experimenting with vacuum-tube circuits patterned after Stibitz's flip-flop.

In June 1941, Mauchly drove to Iowa to visit a young professor at Iowa State University, John V. Atanasoff. At a scientific meeting in 1940, Atanasoff had heard Mauchly deliver a lecture on his weather-forecasting work with the Harmonic Analyzer; Mauchly also touched on the topic of electronic calculating machines, saying they were probably the only way to resolve more complex weather theories. Atanasoff said he, too, had been working with electronic circuits for a calculating machine. His prototype was digital, not analog, and thus was out of the mainstream of university research. Most of the major universities were enmeshed in major experiments on next-generation analog machines. They had grown up on Bush's Differential Analyzer, and their research was devoted to perfecting that technology. Digital was something new and different and not yet accepted.

Mauchly arrived at Atanasoff's house on Friday the thirteenth and spent the weekend engrossed in the gadget, which didn't yet fully work. Then Mauchly received word that he had been accepted into the electronics course at the University of Pennsylvania's Moore School of Electrical Engineering. He hurried back to Philadelphia.

In the summer of 1941, with the war in Europe nearly two years old and Adolf Hitler occupying much of the continent, the United States was reluctantly making preparations to join the hostilities. The U.S. War Department asked Penn to offer a ten-week course called "Defense Training in Electronics," designed to take men from other scientific fields like physics and mathematics and quickly train them in electronics. On many fronts, weapons and warfare were rapidly becoming an electronic endeavor. The Allies had placed a huge premium on scientists in their war effort. The U.S. Army saw Mauchly and the others in the course as needed brainpower; Mauchly simply saw the course as a way to learn more for his weather-forecasting machines. "Here at last was the course that I wanted," he said.

John Mauchly, chairman of the Ursinus College Physics Department, which consisted entirely of John Mauchly, was one of two Ph.D.s taking the course. The oldest student in the class, he was assigned to the youngest lab instructor at the Moore School—Presper Eckert, who had just graduated with a bachelor's degree in the class of 1941.

Pres Eckert: A Genius in His Own Right

John Adam Presper Eckert Jr.—"Pres"—was born a Philadelphia blue blood. His father was a wealthy real estate

Young J. Presper Eckert with his mother and Douglas Fairbanks (From the Eckert family collection)

developer who made his mark building garden-style apartments and high-rise parking garages in the United States and in Europe, where he landed a big contract with American Railway Express, now American Express. John Eckert, nicknamed "Johnny Rusher" because he was so impetuous, had his offices at Fifteenth and Market.

Pres grew up in a large stone house with two chimneys and a carriage house in the Germantown section of Philadelphia, a few doors down from baseball legend Connie Mack. His father moved in circles of celebrity, wealth, and power. He played golf regularly with Ty Cobb and often took his son along on his global travels. The result was pictures of Presper—his grandmother's maiden name—on a camel at the pyramids in Egypt, in a boat cruising the Nile River, and on movie sets with Douglas Fairbanks and Charlie Chaplin. At five, Pres was photographed on a Miami golf outing with President Warren G. Harding.

But Pres Eckert was more than just a child who had been driven to the prestigious William Penn Charter School by a chauffeur. He was a genius in his own right. While other five-year-olds were drawing pictures of rainbows and stick figures, Pres was sketching radios and speakers. At age twelve, he won a Philadelphia science fair with a water-filled tub and a sailboat that he could control with a steering wheel hooked to magnets laid at the bottom of a homemade pond. This invention was patterned after an amusement he had seen in a park in Paris, and it was so sophisticated it had a rheostat that could control electric current to the magnets, enabling him to drop one boat and pick up another for maneuvering in the four-by-six-foot pond. At age fourteen, he replaced a vexatious battery-powered intercom system in one of his father's high-rise apartment buildings with an electrical system. He built radios and phonograph amplifiers, and earned pocket money installing sound systems for schools, nightclubs, and special events. He even was hired by West Laurel Hill Cemetery in Merion to build a music system that masked the unnerving sound of gas burners in the nearby crematorium.

Philadelphia became the center of the young electronics industry in the United States. Philco—Philadelphia Company—was the biggest manufacturer of radios. RCA Victor Company was directly across the Delaware River in Camden, New Jersey. Atwater Kent Manufacturing Company was also based in Philadelphia. The Franklin Institute, established in part by Benjamin Franklin's estate, grew into a major center of research and exhibition. Pres, a member of the downtown Engineers' Club of Philadelphia, drank it all up. In high school, he even spent afternoons hanging out in the Chestnut Hill research laboratory of Philo Taylor Farnsworth, who had demonstrated a working model of a television system in 1927.

On the math portion of the College Board examination, Pres Eckert placed second in the country, behind another Penn Charter student. He wanted to go to the center of U.S. scientific research—MIT—and was easily accepted. But his mother couldn't bear the thought of her only child leaving home, and his father wanted him to attend business school, so they enrolled Pres at the Wharton School of Business at the University of Pennsylvania. Feigning tight finances because of the depression, they even required Pres to live at home and commute to the downtown campus. Bored in business classes, he soon tried to transfer to the physics department, but no spaces were available. A compromise was Penn's Moore School of Electrical Engineering, where he enrolled in 1937.

At Penn, Eckert distinguished himself as a bright young man but not an outstanding student. He was a perfectionist, like his father, orderly and hard driving. But he was not very diligent when it came to classes that bored him, and his grades suffered. "In class, he was always testing the teachers. The rest of us were on the side listening," recalls classmate Jack Davis.

One day Eckert fell asleep in a class taught by Harold Pender, dean of the Moore School. "If you're going to come to class, why can't you stay awake?" Pender frowned at Eckert.

"Why?" Eckert asked indignantly, according to S. Reid Warren, another faculty member.

Eckert made a name for himself in other ways, as well. At one dance, he created the Osculometer—a machine he claimed measured the intensity, the passion, of a kiss. Couples would grab handles wired to the Osculometer, and an array of ten lightbulbs progressively lit up when the pair kissed, completing the electric circuit. What the engineers knew—and their dates didn't—was that if you got your lips wet enough, hands sweaty enough, and held the kiss long enough, you could get all ten bulbs to light up. Then a loudspeaker atop the device would proclaim, "WAH! WAH! WAHHHH!"

In 1940, still only twenty-one years old, Eckert applied for his first patent, which was granted two years later. It was called "Light Modulating Methods and Apparatus," and it amounted to a motion-picture sound system. From mentors at the Franklin Institute, Eckert had picked up the notion of using light beams to communicate sound, and this device recorded sound on film with less distortion than conventional methods by moving a light beam back and forth—a precursor of today's fiber-optic sound systems. The machine never sold, however. The industry stuck with a more mechanical method of etching sound waves on film.

The Moore School proved to be an ideal place for the talents of Eckert, who was just as impulsive as his father. He found himself gravitating naturally to engineering—because engineers get things done—rather than to science, where the search for truth often carries less-immediate gratification. And the Moore School was one of the finest engineering institutions around. It was also one of the few institutions outside MIT to have a Differential Analyzer, the giant contraption of gears and wheels that could solve differential equations. That useful device, plus the school's proximity to several key military installations, helped put the Moore School in the thick of military engineering research. The school was under all kinds of pressure to get things done for the military. For example, faculty members were hired to create a device that could be carried by a plane and generate an electromagnetic field that would detonate water mines and detect submarines. And the Moore School was paired with MIT on major experimentation with radar.

The Moore School (Used by permission of the University of Pennsylvania Archives)

Eckert worked intensely on the problem of accurately measuring time intervals between when a radar signal was emitted and when it bounced back off an object. The elapsed time would determine how far away the object was; thus, measuring it accurately was the key to good radar. The army wanted a device that could measure the bounce-back of the radar signal to a hundredth of a microsecond. Shunning traditional methods, Eckert, who recalled Farnsworth's edict that electronic parts were always better than mechanical, set out to explore the use of electronic counters. He found a circuit, developed by RCA, that could be modified for the radar application, and he paired it with a conventional timing device that used a fluid-filled acoustical tank. A radar pulse was sent out at the speed of light, and a sound pulse was simultaneously reflected back and forth through the tank at the speed of sound. When the radar pulse returned, the distance could be measured by how far the sound pulse had traveled in the tank.

Eckert then took the science of the tank much farther. Drawing on his knowledge of sound recordings, he filled the tank with mercury, which was a better medium for carrying sound waves, and dramatically increased the capacity of the tank by devising an electronic way to repeat the pulses over and over, rather than just echoing them with reflectors. He replaced the reflectors with quartz crystal that picked up the sound wave, then reinforced it and transmitted it by wires back to the starting point. Pulses then moved in only one direction in the tank. You didn't have to worry about collisions that would disrupt the waves. It was analogous, he would suggest later, to reciting something in your head over and over so you don't forget it. Eckert didn't forget; the radar work taught him to build high-speed electronic circuits, and the mercury delay-line tank would later become a crucial piece of computer construction.

The radar work kept the engineers on the roof of the Moore School through a long, hot summer. Most wore shorts and T-shirts and ventured to the air-conditioned

Pres Eckert (Detail; The Smithsonian Institution)

basement, where the Differential Analyzer operated, to cool off and socialize with the women working on ballistic firing tables. "Pres always had on a white linen shirt with a monogram and a black necktie. Always," recalled one of the women, Kathleen Mauchly Antonelli (née Kathleen McNulty), who later became Mauchly's second wife. "I asked him why once, and he said, 'This is what my mother put out for me this morning.' He lived a very pampered life."

Pipe Dreams in the Lab

Though they had different upbringings and were twelve years apart in age, John Mauchly and Pres Eckert became fast friends, wired together by a shared enthusiasm for creating devices. They had amazingly similar childhood interests. Both were fascinated by electricity and wiring, and both had rigged up the same kind of boyhood toys and gimmicks. Eckert was a man more interested in doing than teaching, and prescribed lab exercises bored him. Mauchly knew exactly what he wanted to work on, and saw little value in simple experiments of a caliber he might have assigned to his Ursinus students. Much of the lab time Eckert and Mauchly were assigned to spend together was actually spent talking about different ideas—including computing machines.

"We would sit around on lab tables, dangling our legs, talking," Mauchly recalled. Often their discussions continued over coffee and ice cream sodas at Linton's, a round-the-clock restaurant near the Moore School building at Thirty-third Street and Walnut Street, and the two would spend hours making sketches on napkins.

Would it be possible to make an electronic calculator? Yes, Eckert thought. Difficult, maybe, but not impossible. A machine could be designed to do nothing but count pulses of electrons, with the pulses representing numbers, and to crunch numbers in different ways to solve different problems. Instead of moving gears and wheels in a conventional calculating machine, Mauchly thought he could build a machine with no moving parts; only the electrons would course through the machine. Since they moved very fast, the machine could calculate at very fast speeds and thus solve problems far more complex than could be handled by existing machines. And it would be extremely accurate—assuming it worked.

Reliability was the key, Eckert said from the start. As an undergraduate, he had already designed an electronic device for a chemical company that measured emissions from a smokestack by shooting light through the smoke and calculating the amount of light that passed through. That device provided reliable readings, and it gave Eckert confidence he could go farther—not that Eckert lacked confidence.

Mauchly and Eckert complemented each other perfectly. The lust for big ideas that drove Mauchly from engineering to physics gave the pair many problems to tackle, and the drive to build things that pushed Eckert away from physics to engineering gave the pair the ability to make fuzzy concepts real.

The two began with an idea to use electronics to speed up part of the Differential Analyzer, and they gradually progressed to ways to make all parts of the Differential Analyzer electronic. "We finally came to the conclusion that if you're going to do this, you ought to do it whole hog and make everything in sight digital," Eckert recalled.

With war preparations intensifying, some of the Penn faculty were called to active duty, and Mauchly was hired to fill in with a full teaching schedule at the Moore School. It wasn't that he had dazzled the Penn faculty; he was the only Ph.D. available for the job. Mauchly was seen as undisciplined, an odd bird from a small college who lacked the proper pedigree for an institution that preferred its professors to be inbred. The Moore School was a conservative place firmly grounded in the conventions of electrical engineering. Mauchly was unconventional. Even after he joined the faculty and took on a heavy teaching load, he continued his experiments with circuits and computer design. In the corner of his office, buried under papers, was a prototype for a counting circuit using gas-filled tubes—the kind of tubes used in television sets.

In August 1942, nine months after Pearl Harbor, Mauchly produced a seven-page proposal: "The Use of High-Speed Vacuum Tube Devices for Calculation." Mauchly's idea for a completely electronic machine was rather brazen, and against all convention. Though he touted the advantage that his machine would be far more accurate than existing mechanical devices, his main selling point was speed. "A great gain in the speed of the calculation can be obtained if the devices that are used employ electronic means for the performance of the calculation, because the speed of such devices can be made very much higher than that of any mechanical device," Mauchly wrote in his proposal.

Doing things faster was the order of the day. Henry Ford's assembly line had revolutionized manufacturing, and the pace of American life was moving faster and faster. The

War Department was demanding faster radar, faster construction, faster computations. So were universities. In scientific exploration, much of the low-hanging fruit had been picked. Research had progressed inside the atom and into the far reaches of the atmosphere. Complex problems required machines capable of fast calculations, or academic advancement would be stymied by mathematical headaches.

Realizing he was going against the grain at Penn, Mauchly even tried to make his device sound conventional. He knew the university would not be willing to gamble major funding on a radical project. The new machine, he said, would be "in every sense the electrical analogue of the mechanical adding, multiplying and dividing machines which are now manufactured for ordinary arithmetic purposes." He described the digital system and included the phrase "simple, isn't it?"

At the time, Harvard, MIT, Bell Labs, and other institutions were working furiously on analog machines with moving parts that could add, subtract, multiply, and divide faster than humans. But they were accurate only if every part was in perfect alignment, and, like Penn's Differential Analyzer, they were slow. A digital machine would be more accurate and much faster. But it was a whole new way of thinking, and few were ready yet to abandon analog devices.

Mauchly's memorandum was ignored by Penn's deans, dismissed as the unsophisticated musings of a man known to have a lot of pipe dreams. "None of us had much confidence in Mauchly at that time," said Carl Chambers, a research director at the university. The memo was filed away to be forgotten, and later was declared lost. Most likely, it had been placed in the circular file from the start.

CHAPTER THREE

Crunched by Numbers

As his Studebaker rumbled out of Philadelphia in April 1943, Lieutenant Herman Goldstine couldn't hide his nerves. He was heading to a meeting with colonels and generals and some of the most important scientists in the country, where he would make a pitch for an expensive, top secret U.S. Army project. The young lieutenant should have been enormously excited, but Goldstine was shaking. He was convinced the project was necessary for the Allies to win World War II, and convinced he could make it happen. But he also knew that major scientific advisers from institutions like MIT, no less, had already said it was folly, that the idea would never work. Despite that, Goldstine was betting his reputation, and maybe his career, on the two men he carried in the backseat of his car.

War creates strange bends in the road. After earning a

Ph.D. in mathematics at the University of Chicago in 1936, and having his pick of prestigious faculty positions upon graduation, Goldstine became a professor at the University of Michigan. But then he was drafted in July 1942. He was hardly a warrior. He was a spindly sort of egghead intent on solving the world's mathematical, not political, problems. So it seemed like a blessing when a former professor called a friend with army connections and suggested that Goldstine could best serve his country behind a desk, not a gun. Goldstine was assigned to the Ballistics Research Laboratory at the Aberdeen Proving Ground in Maryland, which was charged with keeping the army supplied with weapons.

Aberdeen is a vast chunk of land jutting into the Chesapeake Bay halfway between Philadelphia and Baltimore. The army used its lush fields and rolling hills to test artillery guns and other weapons. Since a gunner often couldn't see his target over a hill, he relied on a booklet of firing tables to aim the artillery gun. How far the shell traveled depended on a host of variables, from the wind speed and direction to the humidity and temperature and elevation above sea level. Even the temperature of the gunpowder mattered. A gun such as the 155-millimeter "Long Tom" required a firing table with five hundred different sets of conditions. Each new gun, and each new shell, had to have new firing tables, and the calculations were done at Aberdeen based on test-firings and mathematical formulas.

It took more than a month to produce a complete firing table with all the trajectories needed for different elevations, muzzle velocities, air densities, wind speeds, and other variables. At the time, Aberdeen had a team of women doing the calculating, using desk calculators with push buttons and a large handle to pull to complete each arithmetic operation. The women were called "computers."

In early 1943, the war was not going well for the Allies. Hitler was in control in Europe, the U.S. and British armies were struggling in northern Africa, and the U.S. fleet in the Pacific had turned back the Japanese only as far as Guadalcanal. Factories were churning out bigger artillery guns, but that wasn't enough. Aberdeen was falling far behind in its firing table responsibility, and guns were being delivered to Europe and Africa that were essentially useless because they could not be aimed. Different terrain, too, was becoming a huge problem for the army, especially in Africa. Tables made for Europe, the army found, did not work in Africa, where troops landed in the fall of 1942, because the ground was softer and guns behaved differently. The army had contracted with Penn's Moore School for help with all the calculating because the school had a Differential Analyzer. The army also assembled a team of civil-service human computers at Penn to add more number-crunching might. It was still not enough.

Dr. Goldstine, a meticulous mathematician, had been put in charge of the operation at Penn and ordered to get the tables completed faster, no matter what. That seemed simple enough, until Goldstine began calculating the impossibility of the task. The demand for tables was so great, they would never be finished before the guns reached combat. Goldstine sent his wife, Adele, a mathematician herself, on cross-country recruiting trips to seek out more female math majors at colleges, but there was only a handful to be found. He prodded technicians to run the Differential Analyzer as much as possible, but it was prone to breakdowns.

Then one day, a graduate student at Penn asked Goldstine if he had heard of an idea that a newly hired professor, John Mauchly, had been spouting. It seemed so silly that the Penn faculty had ignored it. Mauchly had wanted to make an electronic calculator that could take the place of all the "computers."

An Ingenious Proposal

When Goldstine tracked him down and inquired about his idea, Mauchly couldn't believe his good fortune. He immediately realized the army was the way to get his machine built. Suddenly, John Mauchly became an expert on firing tables. He would venture down to the basement, where the Differential Analyzer was located, and tantalize the firing-table workers with questions like, "Wouldn't it be great if you had a machine that would do that in twenty seconds?"

Captain Herman Goldstine (Detail; The Smithsonian Institution)

"We just thought he was a little bit crazy. He had dreams," recalled one of the women "computers."

"John never set out in life to build a computer," Eckert said years later. "John's goal was to forecast the weather, and it was only incidental that he found out that there was just no computing machine that existed that would handle all this voluminous stuff."

Goldstine was immediately enthusiastic. Only twenty-nine-years old, he was full of hope that new research would yield major breakthroughs. He marched into the office of John Grist Brainerd, the school's liaison with the army, and asked about Mauchly's proposal, but Brainerd couldn't find the six-month-old memo. The memo was re-created from shorthand notes taken by Mauchly's secretary, though the seven-page document had now shrunk to five pages. Brainerd sent the memo on with a faint endorsement: "Read with interest."

Goldstine quickly grasped Mauchly's idea to make the Differential Analyzer electronic—replacing all the gears and wheels with electronic counters driven by pulses of electricity—and he persuaded his immediate superiors to take the idea to top army brass and ask for funding. To stop or start a gear wheel in one-millionth of a second was very difficult, but to start or stop an electron in one-millionth of a second was fairly easy. It made sense to Goldstine; such a machine could be much faster and more accurate.

Penn didn't take Goldstine any more seriously than Mauchly. "Dr. Brainerd was pooh-poohing the idea that anybody would ever seriously consider such a thing," Mauchly recalled in a diary entry made a year later, in 1944. "He told me that Goldstine was a very young fellow, just a kid, really, who hadn't been around, and not to pay much attention to his statement that there was a good chance for the thing to be put through."

But Goldstine had gotten the idea far enough to be in his Studebaker on the way to a fateful meeting in Aberdeen to meet with the top brass. Goldstine drove because the army gave him generous gas-rationing points. (Gasoline was in scarce supply in Philadelphia because most of the nation's fuel was directed to the war effort.) In the backseat of the Studebaker, he carried the unlikely tandem of Mauchly and Eckert. And in the front passenger seat sat a rather skeptical Dr. Brainerd.

What Herman Goldstine would present on April 9, 1943, were two nobodies—a kid and a dreamer. Eckert turned twenty-four years old that very day. Even Goldstine wondered how he could get the army to take them seriously. They had what seemed like a great idea, but here they were, driving to the most important meeting of their lives, and Mauchly and Eckert were in the backseat with paper and pencils frantically rewriting sections of their proposal to the army.

These were desperate times, so desperate, the army was willing to listen to any harebrained idea brought in by a brand-new lieutenant. German U-boats were sinking ships, even near the U.S. shores of the Atlantic Ocean and the Gulf of Mexico, and fighting in Africa was fierce. "At that time during the war, people were looking for anybody with new ideas," recalled Lila Butler, one of the human computers working on the firing tables. Even Mauchly knew the war was the only reason his wacky proposal was getting a second chance after being trashed by Penn. "The war got worse," he said. "When I first presented the idea, we weren't at the same stage of desperation."

Once at the army base, Mauchly and Eckert were put in a room with a secretary to make final changes while Brainerd and the others went to the officers club for lunch. Mauchly, who had to eat regularly because of a genetic condition that caused anemia, was beginning to feel sicker and sicker.

By their nature, army officers are a skeptical bunch. Soldiers know it pays to be paranoid and untrusting, and the leadership at Aberdeen was considered a tough sell. Not long before Goldstine was to make his pitch, another group of scientists had gone to Aberdeen to tell the army about the newly discovered science of nuclear fission—the splitting of the atom—and how they could use it to make a bomb of enormous power. Aberdeen said no to the proposal, only to be overruled by President Franklin D. Roosevelt himself. Indeed, at the first meeting of the Advisory Committee on Uranium, held shortly after Hitler had invaded Poland, Lieutenant Colonel Keith F. Adamson, the army's ordnance expert, was still sneering at the notion. "In Aberdeen," he said at the meeting, according to an account by committee member Edward Teller, "we have a goat tethered to a stick with a ten-foot rope, and we have promised a big prize to anyone who can kill the goat with a death ray."

The Manhattan Project, code-named Project Y, had been born in the fall of 1942, and now a few months later, Goldstine was marching into Aberdeen with another huge, speculative scientific project.

"Simon, Give Goldstine the Money"

The meeting included Colonel Leslie E. Simon, director of the Ballistics Research Laboratory, and Oswald Veblen, a renowned mathematician and technical adviser to the army laboratory. (Veblen and Albert Einstein were the first professors appointed to the Institute of Advanced Study at Princeton University.) It was a group with tremendous academic credentials, not merely a collection of army officers.

Goldstine expected a tough fight, but he didn't get it. Instead of the skepticism that the army had shown the earlier scientific pipe dream, this effort was more easily embraced by army brass, perhaps only because it was cheaper and scientifically less bewildering, or perhaps because they had been overruled the last time they scoffed at a wild new idea. Electrons were widely understood, a part of everyday life that powered radio, radar, telephones, and the emerging field of television. And what Mauchly and Eckert had proposed was pitched not as new science but rather as a conversion of existing technology to electronics. They proposed a machine that would, in essence, string together a dozen desk calculators and allow them to "talk" to each other, so the results of one calculation could be used in another to solve complex problems. It was to be a flexible, general-purpose machine, capable of not only crunching firing tables but also performing wind tunnel calculations and maybe even weather-pattern predictions. And it certainly wasn't as bizarre a proposal as building a bomb that would shoot a plug of uranium into a chunk of the same material and create a city-leveling explosion.

Once the meeting finally got under way, Dr. Veblen listened for a short while and then cut Lieutenant Goldstine off. Tipping his chair back on two legs, Veblen delivered his pronouncement: "Simon, give Goldstine the money."

That was it. Project PX was born.

The army gave the University of Pennsylvania a development contract, no. W-670-ORD-4926, and an initial appropriation of $61,700 for the first six months of work on what Eckert and Mauchly called the Electronic Numerical Integrator. The name was later changed to Electronic Numerical Integrator and Computer at the suggestion of Colonel Paul N. Gillon, the assistant director of the Ballistics Research Laboratory. In fine army fashion, the MIT-educated Gillon had given the project a catchy acronym: ENIAC.

Decades later, twenty-somethings and teenagers would change the face of computing with garage tinkering. But in the 1940s, it was amazing that so much hope and public financing were invested in such young bucks. Mauchly was only thirty-five, and he was the wise old man of the bunch. "I don't think anybody would have given a young kid twenty-four years old all this money to do something if there hadn't been a war on," Eckert once said.

Still, youth has its advantages. "Had John and I been five years older and that much more experienced," he added, "we might have 'known' a true electronic computer would not be built."

CHAPTER FOUR

Getting Started

Despite their excitement and impatience, Eckert and Mauchly started work cautiously. They knew something that none of their benefactors did: Although they had the basic outline of what they were going to build, they had not worked out exactly how they were going to create it. And unlike Aiken, who had all those IBM engineers working to make his concept a reality, they were on their own.

ENIAC, they had decided, would have three main parts. First, there would be self-contained machines to handle the math operations: units built for addition, a high-speed multiplier, and a box wired to handle division and square roots. Second, there would be memory units to store numbers and instructions. Most of these would be electronic, since numbers had to move quickly through the machine. But some would be big mechanical panels with switches that could be set to represent constants in a calculation. The numbers drawn from these panels would be electronic, but the values would be set with switches. Either way, there would be no paper tape, because anything on paper was too slow. Punch cards would be used only in the initial phase of inputting the original problem. Third, ENIAC had to have a master programmer to control the machine, not to actually crunch numbers but to bark orders to the rest of the machine and keep its electron pulses in order. (Terms like "memory" and "programmer" were already becoming standard computer terms.) In addition to the three main parts, ENIAC needed some peripheral controls, like a unit to initiate the computation and a cycling unit to keep it all synchronized. ENIAC would be like a forty-piece orchestra, and it had to have a strong conductor.

This is the basic structure of computers even today.

The different units were wired together so that numbers could be shared and instructions routed over cables. Mauchly envisioned cables bundled together in phonelike trunks he and Eckert called digit trays, and similar cables bundled into program trays. Each digit tray would have eleven wires so it could carry a ten-digit number and a plus or minus sign. Digit trays were permanently wired throughout the machine. Each program tray also would have eleven wires, but it could be moved around and plugged in to send signals to the entire machine or to specific units for special instructions. Program trays would be the electronic cousin of the shafts that made up the veins and arteries of Bush's Differential Analyzer or the cranks that were turned by Leibniz and Babbage.

But Mauchly's scheme wasn't based on a careful reading of the history of computing machines. It was simply based on his desire to string together ten or twenty desk calculators, like the kind he had used for his weather-pattern computations. He borrowed some notions from the Differential Analyzer, but his real inspiration was to make the desk calculator electronic. And by stringing many calculations together in a series, the machine could handle complicated problems. Mauchly, indeed most everyone associated with ENIAC in its infancy, had never heard of Babbage and his design because they were outside the mainstream of computing machine research. So they had no notion of the concept of a central processing unit—Babbage's "mill"—where the actual number crunching would take place. ENIAC's computing power would be decentralized, spread out through many boxes, including twenty accumulators that could be used both to store numbers and to add them together. ("Accumulator" was a term used in desk calculators.)

Together, Eckert and Mauchly started by exploring issues in the logical design of the machine and what materials they wanted to use. The basic science of getting circuits to count already had been developed—the Stibitz flip-flop design—but none of the existing designs of counting circuits would work with ENIAC. Eckert was confident he could design counting circuits for the machine.

A trickier problem was devising a means to control the circuits and be certain that the right pulses got to the right circuits, and that the circuits counted in the right order at the right time. Initially, control—the programming of the machine—was a leap of faith. Eckert and Mauchly believed they would be able to devise a control, but they didn't know how they were going to do it.

"The smartest thing we did was to start slowly," Eckert said.

Assembling the ENIAC Team

They began in July 1943 with only twelve people from the faculty assigned to the project, which was relegated to a drab first-floor room with darkened walls and open rafters in the Moore School building, a former musical instrument factory. None of the senior faculty engineers took an active role. It was clear the Moore School still had little faith in, and not much enthusiasm for, the endeavor.

Youth worked to the project's advantage, however. The group assigned to ENIAC was an odd yet powerful collection of engineering talent. It included Bob Shaw, an engineer beloved for his practical jokes and inventive designs; Arthur Burks, a Ph.D. philosopher who had taken the defense electronics training course with Mauchly and, like Mauchly, had been offered a faculty position; Kite Sharpless, a logical, intelligent, and considerate engineer from a Philadelphia Quaker family; Chuan Chu, a high-born Mandarin Chinese immigrant; and Jack Davis, a happy-go-lucky bachelor who speculated in grain futures with Bob Shaw just for fun.

Each was given a specific piece of the machine to engineer, based on Mauchly's concept and Eckert's layout. Not a single schematic was drawn without Eckert's imprint, however.

Shaw stood out as a great motivator, despite suffering from albinism and a spinal disease that left him frail, hobbled, and nearly blind. He worked with his nose almost touching the paper, drawing very large circuits on three-by-five-foot sheets, even though he could see only a six-inch square at a time. He worked hard, never complained about his health, and was a source of merriment among the group. "He spoke well, wrote beautifully, was a marvelous teacher, had a sly, kinky sense of humor," said Jean Bartik, one of the women doing ballistic calculations in the basement of the Moore School. "He was always ready for a party and a good time."

The group was spurred on by the excitement of a new project and the pressure of the war. The Allies were beginning to advance against Hitler and Benito Mussolini. Allied troops had invaded Sicily in July 1943, just as the ENIAC project started, and the U.S. Air Force was bombing Rome

ENIAC's design team, from left to right: James Cummings, Kite Sharpless, Joseph Chedaker, Bob Shaw, John Davis, Chuan Chu, Harry Huskey, Pres Eckert, Captain Herman Goldstine, Arthur Burks, Brad Sheppard, F. Robert Michaels, and John Mauchly (The Charles Babbage Institute)

in advance of an invasion. The need for firing tables was at a fever pitch, and the ENIAC project operated every hour of the day and night. Engineers came and went at odd hours. Eckert and Mauchly were both night owls who worked nearly around the clock, usually arriving in the late morning and leaving in the early morning. The schedule called for everyone to work seven days a week, and every Saturday morning, there was a meeting to talk out problems and update university officials. Carl Chambers, one of the

Moore School's three research directors, always attended. John Grist Brainerd, who had been appointed chief administrator of the ENIAC project despite having sneered at the idea initially, also came to the meetings, "but he wasn't much up on what was going on," according to Jack Davis.

Lieutenant Goldstine, the army's point man on the project, was an almost constant companion to Eckert and Mauchly, often joining the pair for three meals a day. Goldstine was more than just a supply sergeant for the group. He was keenly involved in the design and construction of the machine, frequently meeting with the engineers late into the night to discuss the logical design of circuits and the mathematical methods of computation. "He was the lubricant that made that thing work," Eckert once said of Goldstine.

Eckert now had a constant companion at home, for he had gotten married that year. His bride, Hester Caldwell, was a member of Philadelphia's prominent Wallander family. Eckert had met Hester on a train when she was engaged to someone else; he later persuaded her to marry him instead.

The First Computer Takes Shape

The group started working on the accumulators first. Eckert and Mauchly wanted them to be able to store a ten-digit

Close-up of an ENIAC accumulator (The Hagley Museum and Library)

number with a positive or negative sign. Instead of simply having the accumulator count pulses to reach a number, they split each digit of a number into its own circuit. So instead of the number 333, for example, requiring 333 pulses, it required 3 pulses in the hundreds circuit, 3 in the tens circuit, and 3 in the ones circuit—for a total of just 9 pulses.

That design amounted to a huge savings when it came to eight- and ten-digit numbers, and greatly enhanced the speed of the machine, even though each pulse took only two-millionths of a second to transmit. Speed was what they had promised the army; hence anything that increased speed was welcomed. To avoid other data-slowing bottlenecks, Eckert and Mauchly decided the accumulators not only had to store numbers but also had to be able to add and subtract them and transmit the result elsewhere within the machine.

Counting circuits proved more difficult to construct than was first thought. Although four counting-circuit designs were known to Eckert and Mauchly, they had all been designed for another purpose, such as physics experiments. "It was hoped originally that counter circuits and other circuits of basic importance would be available from the work of others," the first progress report on ENIAC noted. "Actually, this has not been found to be the case, and a large amount of effort has been given over to the development of circuits of sufficient scope and reliability to be included in ENIAC."

Instead of the electromechanical relays that Stibitz used in his flip-flops, Mauchly wanted to employ vacuum tubes, which he had seen used for counting in physics experiments at Swarthmore, and which he had experimented with in his own early contrivances. (Eckert had worked with vacuum tubes on radar projects.) Relays and vacuum tubes operate in much the same way: The simplest vacuum tube has two states, on or off, just as the relay has two states, open or closed. But electricity flows through wires at the speed of light, and the armatures of relays can't move nearly that fast. The vacuum tube could toggle a lot faster than the relay.

A simple vacuum tube has a cathode, which gives off

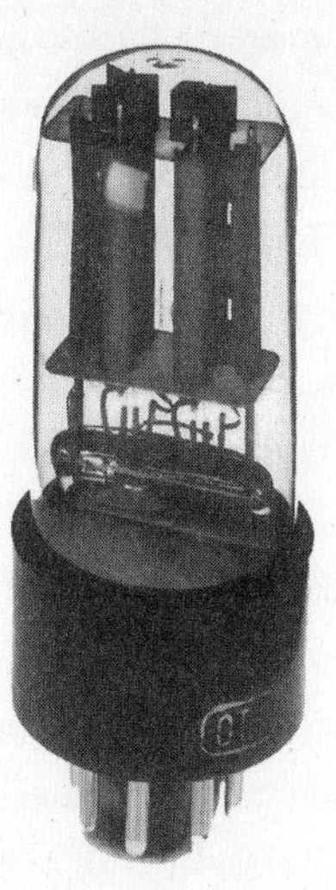

A vacuum tube similar to the kind used in ENIAC (Used by permission of Bill Westheimer)

electrons when stimulated, a plate that receives the electrons, and a grid that controls the current passing through. There's also a heater to bring the cathode to the temperature required to emit electrons. If the voltage sent to the tube is high enough, it turns on. The higher the voltage, the more electrons are emitted. Vacuum tubes were developed for their ability to modulate and amplify electric current. In a television set, for example, the picture can be controlled with vacuum tubes. In ENIAC, they were used only as on-off switches.

Eckert and Mauchly studied the work of Perry Crawford at MIT, who had designed an electronic control for antiaircraft guns and published a master's thesis on counting circuits. They also visited Radio Corporation of America (RCA), which manufactured tubes and circuits. The trip was arranged by Goldstine, and Eckert and Mauchly worked both with tube engineers and with the company's renowned scientist Jan Rajchman in trying to modify precision timing circuits for use as electronic computing circuits.

In fact, RCA had been invited by the Moore School to participate as the primary subcontractor on the project, but declined. This fateful decision has been attributed to Vladimir Zworykin, director of electronic research and a coinventor of television. Zworykin thought it impossible to get that many tubes working together, and RCA's scientists thought the Moore School youths were "extremely naive," though quite enthusiastic. Nevertheless, RCA agreed to provide technical advice in the spirit of "wartime cooperation." The company had been offered the chance to launch the computer industry, and it passed.

Finally, Eckert hit on a design of a counting circuit that would work in a computer. The invention satisfied all his needs: It could add or subtract and carry numbers when the sum was greater than 9. It was fast and foolproof, and it could transmit a result to the next unit. The counter would be a circuit of ten flip-flops—each flip-flop was two vacuum tubes—arranged to form what they called a decade-counter ring. Eckert wired the flip-flops so that only one tube could be on at a time, a crucial bit of engineering that greatly eliminated errors. If tube A was on and the next pulse came in, the flip-flop flipped to B on, A off, and sent a pulse to the next flip-flop in the circuit. If the fifth flip-flop, representing the numbers, was on, and a series of three pulses came in because the computer wanted to add 3 to 5, flip-flop five would be turned off and the pulses would ultimately come to rest with flip-flop eight—and only flip-flop eight—turned on. That circuit would represent the sum of 5 plus 3. In all, a decade counter had to have twenty tubes for each ten-digit number, and eight other tubes for the plus or minus sign and other controls.

Mauchly was aware of counting devices that worked in base 2, where numbers were amalgamations of 1s and 0s seemingly tailored for the language of the flip-flop. But he and Eckert decided their machine best stick to the more familiar base 10 of the decimal system. Not only did they want ENIAC to be as "conventional" as possible, but they also figured using base 2 would actually require more vacuum tubes than base 10. Yet one by-product of the decision was that the counting circuits were more complicated than Stibitz's design.

The decade counter provided the crucial technology to turn pulses into numbers, giving order to the electrons racing inside the machine. The counter was essentially an electronic version of the wheel with ten notches that would rotate as it counted, as Pascal had envisioned. The circuits were self-contained units, wired into a steel chassis in a modular design. If one circuit became suspect, it could be pulled out and replaced with a spare without having to search out exactly which tube had failed. At one point, the team even called the invention the Eckert counter.

"Although we had high hopes at the start, we were far from certain that the computer would become a reality. However, at the end of three months, we knew it would," Eckert said. "We built many different counting circuits, finally finding one that demonstrated the large margin of safety needed for our machine. This was the first tangible evidence that we could complete our job."

ENIAC's accumulators ended up as nine-foot-tall monsters jam-packed with so many tubes that from a distance their back sides looked like the points of a Jumbotron giant television screen. The tubes were socketed in a mesh of wiring that caused lights to blink on the outside panel. The lights were put on the outside to make it easy to keep track of which circuits were functioning; undoubtedly, this feature led Hollywood filmmakers to believe that computers must have blinking lights to be working. Opening up the accumulators while ENIAC was powered up was so dangerous, Eckert designed safety switches into the door panels that would shut the voltage down if opened, so no one would be electrocuted.

Looming over the entire project was a huge question—maybe the biggest problem they faced: Could the team ever

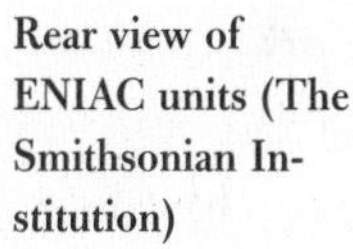

Rear view of ENIAC units (The Smithsonian Institution)

get all the vacuum tubes to work? Vacuum tubes were notoriously unreliable because they burned out easily. If one tube burned out, it would throw off the whole calculation. ENIAC was originally designed to have 5,000 of these fickle creatures, which were supposed to operate at 100,000 pulses per second. This meant that each second there would be 500 million—*half a billion*—chances to screw up.

Many eminent scientists of the day, including those at MIT, doubted ENIAC could ever run accurately because it relied on so many unreliable vacuum tubes. Television sets typically had only thirty tubes, and they often needed repair. An electronic organ known as the novachord, which was built in the late 1930s, contained about 160 tubes. And the biggest use of vacuum tubes at the time was a couple hundred in a counter built at Los Alamos for the Manhattan Project. ENIAC was supposed to have more than twenty times as many. Daunting as that might seem, the problem was even worse because tube quality was declining as the war drained skilled labor from factories. "If it was going to work, one had to be one hundred times more careful," Eckert said.

Eckert had done the math, and he knew the challenge he faced. He tested operating tubes in different ways to hit on the most reliable structure. He searched for the best tubes he could find, and wound up with some devices the telephone company had built for a transatlantic cable. And he found that if he reduced the voltage below what the tube was designed for, he could prolong its life. He ended up running tubes at less than 10 percent of the standard voltage for which they were intended.

"All of our circuit designs were checked algebraically and by detailed calculations. We also were very cautious. If a manufacturer claimed his product had a certain tolerance, we allowed for less," Eckert said.

His paranoia about malfunction was apparent in all aspects of the design. Eckert acquired some mice in cages and starved them a few days. Then he put different kinds of wire in their cages to see which kind they enjoyed eating. The least appetizing brand was used in ENIAC. And since

ENIAC would have 4,000 to 5,000 knobs, Eckert designed his own knobs around specially made screws that were tapered and hardened to minimize the chances they would get loose.

"People thought I was a nut being so fussy about these standards," he said.

Goldstine continued bringing in outside help. Moonlighting telephone company workers, who didn't have much regular work since the war was on and few companies were opening new offices, were hired to do much of the wiring inside the machine. Miles of wiring made up not only the counting circuits and control features but also the digit trays and program trays. Putting all that together was enormously complicated. If any circuit was wired wrong, ENIAC would produce incorrect answers. All it took was one misplaced connection.

Other prominent scientific organizations were consulted. Bell Labs was called in for aid because some of the machine was made up of telephone parts. IBM designed a special card-reading machine to load numbers into ENIAC and spit out the results. An outside engineering firm was hired to design a twenty-four-horsepower ventilating system to draw off the enormous heat generated by the thousands of tubes.

Once the design of the accumulators was well in hand, attention turned toward the "master programmer" that orchestrated certain operations and loops. The master programmer would make subroutines possible by telling the machine to do a routine until something happens, such as a result turning from positive to negative, or repeating a routine one hundred times or until one component is greater than something else. For ballistic shells, for example, the computer should stop calculating trajectories once the shell reaches its target. So when the altitude of the shell falls to zero, the computer moves on to something else. The master programmer would make the machine automatic, and very fast. And it would make it fundamentally different from a calculator: The computer can *do* something with the numbers it produces.

Computers have the ability to react to data, and that stems from the notion of a subroutine, the "If . . . then" statement in programming. If X happens, then do this. If Y happens, do something else. Babbage had the idea of a subroutine a century earlier, but Mauchly and others involved in ENIAC would say in later interviews that they were unaware of Babbage, and that Mauchly came up with the subroutine idea for ENIAC on his own.

Mauchly knew that the key to the machine was "the control of these things"—the different functions. ENIAC had so many elements, it had to have good control. The brilliant design, then, was born of simple practicality.

Mauchly and Eckert did not have control over everything, however. The army's scientific leadership in Aberdeen, where Goldstine's boss, Colonel Paul N. Gillon, resided, inexplicably kept expanding the scope of the program. Those ballooning ambitions forced the design to grow larger and larger to handle the more complex challenges that Aberdeen wanted to put to ENIAC. Instead of 5,000 vacuum tubes, ENIAC grew to about 18,000. Instead of ten accumulators, there would have to be twenty.

And the design of how programming would be handled changed as well. Each accumulator would have to have more programming capabilities built into it. Rather than just adding machines, they would become more like tiny computers themselves strung together. About 30 percent of the parts inside the accumulators ended up being programming equipment. That way, the machine could give accumulators signals for operations that could be carried on inside the accumulator, and operators could "program" ENIAC simply by setting different switches, such as a knob on the outside of an accumulator to tell it whether to perform addition or subtraction.

The back room of the Moore School turned over to the project became a beehive, and it took on the irreverent nickname of the Whistle Factory. Each engineer had a worktable up against a wall of the room. Assemblers and wire men occupied the central floor space, where the machine was taking shape. The pace grew even more relentless. The engineers became the prototypes for the original Silicon Valley nerds. Often they worked around the clock;

sometimes they could be seen racing across the bridge over the Schuylkill River in Philadelphia to catch the last streetcar of the night. "A serious attitude prevailed," recalled Herman Lukoff, one of the engineers. "Everyone realized the importance of his job."

The Team Leaders: Contrasting Styles and Personalities

By the fall of 1943, the Germans were occupying Rome, and the Allies, having moved through northern Africa, were scratching for any foothold in Italy. Most of Europe was a German fortress. Near Naples, U.S., British, French, Polish, and Brazilian troops battled veteran German troops, who, aided by their mountainous perches, proved almost incapable of being removed from southern Italy. Then, in January, the Allies landed 70,000 troops and 18,000 vehicles on the beaches of Anzio, behind the lines of the Third Reich and only about thirty miles from Rome. The Germans controlled the high ground, and the battles were horrifically bloody. Troops established a beachhead but bogged down in mud and slime and moved no farther inland.

Pressure on the ENIAC project grew, and the work in Philadelphia became more and more intense. Still, few strains developed within the core of the Project PX team. Eckert and Mauchly had such totally different personalities they left many bewildered at how they could get along so well.

Pres Eckert was a whirlwind of quick, nervous energy. The balding youth was described as a fidgety-widgety kind of guy, a fingernail biter, and his habits soon became legendary among the ENIAC staff. Eckert carried a pocket-watch chain with no watch, and when he really got wound up, he would twirl the chain endlessly. He would assign a task, explain exactly how he wanted it done, and then correct the work once completed. When problems arose, Eckert always came to the rescue. "I was scared to death of him," recalled Jean Bartik. "He had a fiery temper that could make grown men speechless."

Among the workers, Eckert was constantly looking over his shoulders. "There wasn't a single one of the staff . . . that he didn't tell him where to solder the joint," Carl Chambers said. Some mornings, Eckert came in and told engineers to tear up what they had done because he had thought of a better idea overnight.

Yet the team had unwavering respect for Eckert, and genuine fondness for him because he was not an angry person, just temperamental. Mistakes that brought a tongue-lashing from the cantankerous Eckert were quickly forgotten, and a more charming Pres Eckert would make amends. "Eckert set the most exigent standards and insisted that there be no exceptions," Goldstine wrote in his history of computers. "Eckert's standards were the highest, his energies almost limitless, his ingenuity remarkable and his intelligence extraordinary. From start to finish it was he who gave the project its integrity and ensured its success."

Failures excited Eckert as much as successes, because knowledge was gained by knowing what didn't work. When it came to ideas, Pres Eckert bit into them with full force and didn't let go until he had chewed every morsel of substance from them. His powers of concentration were enormous. "Eckert was always pressing everything to the limit," recalled Jack Davis, Eckert's former classmate who became an engineer on the project. Eckert would glance at a design after not looking at it for several weeks, and immediately zero in on a flaw. He would drop into the lab in the evening to check on something and end up staying most of the night. "He'd get an idea and pursue it relentlessly, call people on the telephone, and talk for an hour," Bartik remembered. "Sometimes he'd start talking to someone, walk out of the building and down the street without ever realizing what he was doing."

John Mauchly, on the other hand, was the calming influence, the lovable intellectual who charmed and always kept an eye on the big picture. Whereas his partner was completely work oriented, Mauchly was more people oriented. A compulsive chain-smoker, he could spontaneously deliver delightful lectures about *Alice in Wonderland* or quote lines from the latest Broadway tune or discuss the most recent news item. "He loved to talk and seemed to develop many of his ideas in the give-and-take of conversations," recalled Bartik. "John enjoyed social occasions, liked to eat good food and drink good liquor. He liked women, attractive young people, the intelligent and unusual. He was primarily an intellectual. . . . Although he could be cynical, he was basically the optimist, expecting the rational to prevail."

Despite his cerebral reputation, Mauchly was intimately involved in the engineering of the machine, and Eckert sought out advice from Mauchly alone. Once when Mauchly was bedridden with the flu, Eckert took questions to Mauchly at his home and came back with a decision on a crucial design. To those who didn't know the two well, Eckert's loyalty to Mauchly was puzzling: Why did he need Mauchly when he clearly was a superior electrical engineer? Mauchly, who maintained a full teaching load throughout the first year of the project, could jump in wherever he was needed. For example, when Kite Sharpless decided to go skiing, he simply left a note in his mailbox saying, "I'll be back in a week." Sharpless had been designing the central timing unit in the master programmer, the metronome of the machine that kept it all synchronized. Without that, no other components could be run together. "We were really stuck. Without it, we couldn't test anything else," said Eckert. Mauchly spent extra time doing the design work, taking over the electrical engineering, soldering iron in hand, even though he was not an electrical engineer.

"John had brilliant ideas that came up suddenly, which he would work on like hell for weeks at a time and then do nothing at all related to the computer for a few weeks while he tried to catch up on his teaching or something," Eckert said in an interview.

Mauchly was also renowned for being perpetually distracted. His diary includes an entry about driving with Eckert, running out of gas, and simply abandoning the car and taking a streetcar. One time at home he noticed a spot where the wallpaper had come loose and deduced that it must have been caused by water from a leaking roof. "Guess this is a job for a roofer," he remarked. The next day, Mauchly asked his wife when the roofer was coming. She told him she thought he would take care of that. "Well, it was my idea," he said. "Why can't you carry it out?" It was a pattern repeated over and over. Mauchly loved generating ideas but hated the work of following through on them. "Why do you think we built a computer?" Eckert once joked.

It was Goldstine who picked up the administrative duties that Mauchly couldn't have handled. Goldstine described himself as a "natural administrator" who knew how to handle groups. He also knew how to handle personalities and sometimes had to help Eckert control the impulsiveness of Mauchly. "Mauchly was a physicist. He was interested in building whatever the hell would work in the shortest period of time," Goldstine said. "An engineer builds something so it can be manufactured and so that it lasts. A physicist may use something only once, so he builds it out of any crap he can get his hands on. It could have been built out of ladies' hairpins. He just wanted something built. But Eckert had very rigorous standards. They complemented each other very well."

The teamwork between dreamer Mauchly and administrator Goldstine wasn't as smooth. "I didn't get along as well with Mauchly as Eckert did," Goldstine said in an interview. "I'm sort of dramatic. Mauchly was not dramatic. His work wasn't done by the numbers, and I like it done by the numbers. He was a nice guy, very smart, but probably was a space case."

Mauchly, in fact, was deeply insecure about what his role really was. Eckert was chief engineer, a title with which no one quibbled. Mauchly, on the other hand, was sometimes listed as consultant or research engineer—titles hardly in keeping with his role as the originator of the idea.

Anxiety and deep soul-searching filled his diary, even as early as the fall of 1944. "Eckert, in recent conversations, seems to indicate that he values my presence among the PX staff members," Mauchly wrote. "He thinks that one difficulty may be that his contributions are tangible, and that mine are more intangible, so that others (such as Brainerd and Pender) are not fully aware of what I have contributed."

Judging by Mauchly's salary, one would deduce he was a rather minor player on the ENIAC team. It wasn't until a year into the project, in 1944, that Mauchly was allowed to work on it full-time, and with that change, the Moore School administration actually forced him to give up one-third of his salary, which was cut to $3,900 a year from $5,800. "Being cut back one-third upset the domestic economy, of course, and that was bad enough. But the implications were that I wasn't worth very much to the project which I had originated," Mauchly wrote in his diary.

In June of 1944, when Eckert and Mauchly completed the first two accumulators and moved the ENIAC project into high gear, Mauchly obtained a part-time consulting job doing statistical work at the Naval Ordnance Laboratory to compensate for the salary cut. The lab was run by John V. Atanasoff, the fellow gadget maker whom Mauchly had visited in Iowa three years earlier to see his computing device. Mauchly spent one day a week in Washington at Atanasoff's lab.

Despite the anxieties, uncertainties, and vanities among officials at Penn and Aberdeen, the core of the group hung together, pulled in by the gravity of the project. When things wouldn't work, frustrated workers referred to the machine as the MANIAC. Harry Huskey, another of the engineers, remembered joking that if they gave their drawings to the Germans, they would set back the war effort ten years. "We were young and deeply involved. We felt like the whole war program depended on us," Goldstine said. "There was a real sense we were doing something very extraordinary."

CHAPTER FIVE

Five Times One Thousand

By February 1944, the ENIAC team had final wiring diagrams and blueprints done, and construction of complete panels was begun. Yet thoughts had already turned to the *next* machine. Eckert and Mauchly had realized the shortcomings and inefficiencies in ENIAC's first design but knew they had to go with it if the wartime project was ever to be completed rapidly enough to be of military use. Still, Eckert was composing memos for a more advanced computer. In January 1944, he had laid out a scheme for a bigger memory, one that could store a full "program" inside the computer and eliminate many of the switches, accumulators, and cabling in ENIAC.

Having been pinned down for months, Allied troops were now advancing in Italy, and preparations were under way for a major offensive in France. The Allies were slowly scoring victories, but at a heavy price. And the demands for better guns and new firing tables were, naturally, ever more urgent. Anything to gain an edge on German defenses.

By mid-June, just days after Allied troops invaded at Normandy and General Dwight Eisenhower declared "the tide has turned," the Philadelphia engineers got the first units of ENIAC to work. It was only a year after the project had begun. They had two complete accumulators containing more than 1,000 tubes—or about one-fifteenth of the final device. They fed some simple problems into the accumulators, and the units solved them. Eckert and Mauchly even patched the two accumulators together to work more complex differential equations, such as producing an exponential function, a parabola, and a sine function. Each time, the machine worked.

"We have finally done it!" they shouted to coworkers, running from the room.

Eckert and Mauchly triumphantly summoned two women "computers" from the basement, where they were working with the Differential Analyzer, and invited them into their secret, usually padlocked, first-floor catacomb. One accumulator was loaded with the number 5, then the other with the number 1,000, and the two were multiplied together. The machine showed 5,000. "I was astounded that it took all this equipment to multiply 5 by 1,000," recalled Kathleen Mauchly Antonelli, one of the computers who witnessed this feat.

But the result had more significance for others. It demonstrated to the army, to the scientific community, and even to skeptics within the Moore School that an electronic computer could work. "We saw answers glowing on 204 small neon lamps," Eckert said. "This was enough to convince the necessary few that an electronic computer was being born."

That summer, with the rest of the units under construction, Goldstine scrambled to find necessary supplies, since the war effort swallowed up anything used to manufacture electronics and even the steel needed to fabricate cabinets for ENIAC. The mathematician proved to be a tireless supply master, however, impressing upon suppliers the fact that the project he was working for was itself a top secret part of the war effort. In reality, ENIAC never got much help from army superiors. Eckert even had to fight against being drafted while work on ENIAC was in full swing. Six times he was summoned in front of a draft board until letters were secured from the head of Civil Service and the head of Ordnance.

Eckert and Mauchly were like mother hens, nursing the machine every hour they could, chasing bad solders and bad tubes, battling to get the thing perfected. With more tubes running, the heat in the musty room became oppressive, and everyone worked in their undershirts.

One night, when Eckert was asleep on a cot beside the machine, two technicians picked up the bed and carried it to an elevator, then placed Eckert in an identical, but empty, room on the second floor. Eckert awoke and was startled to find his machine "stolen"—until it dawned on him what had happened. Such pranks helped everyone cope with the frustrations and setbacks. For example, there was a near catastrophe when a cooling fan in the bottom of a panel caught fire, and flames shot straight up. Luckily, the damage was confined to that single unit. "Neighboring units did not catch fire. If that had happened, it would have been a total disaster," Goldstine said.

In late August, Eisenhower-led Allied forces liberated Paris, British troops closed in on Brussels, and U.S. forces routed the Japanese from Guam. It was beginning to look as if the war might be won before the machine was completed. Although this development took some of the pressure off the ENIAC team, they moved ahead at full steam and began to train workers to run the beast.

"A Son-of-a-Bitch to Program"

There was nothing simple about the training. Complex problems were broken down ahead of time into a series of arithmetical operations. Programmers had to plug units together in the proper order for the flow of the problem, and plan for where and when numbers would be stored. To add 1,234,567 and 2,345,678, for example, programmers would store the first number in accumulator one, the second in accumulator two, and tell accumulator one to send its number into accumulator two, which was programmed to combine pulses. Accumulator two then would register 3,580,245. If the problem called for subtraction, the sign of one number would be reversed and then it would be sent to the other accumulator. If it was a simple multiplication, like 5 times 1,000, the accumulators could be set up to do multiple additions. One thousand could be added together five times.

ENIAC had special units for high-speed multiplication and division. The high-speed multiplier worked much like a human would. It had a built-in multiplication table (7 times 1 equals 7, 7 times 2 equals 14, 7 times 3 equals 21 . . .), so it could break the problem down into simple multiplication of each digit; the result would be sent to an accumulator, eventually tallying the final product. The multiplication process for two ten-digit numbers took all of 2.6 milliseconds. The high-speed divider and square-root unit worked by repeated subtraction and addition. If 738 were to be divided by 6, 6 would be subtracted from 7, with remainder 1. Six would then be subtracted from 1, but that yielded a negative number, so the unit would advance to the tens place, carrying the remainder (1) from the hundreds place. Next, six would be subtracted from 13, leaving 7. Six would be subtracted a second time, leaving 1, and a third time, turning negative. The number 2 would then register

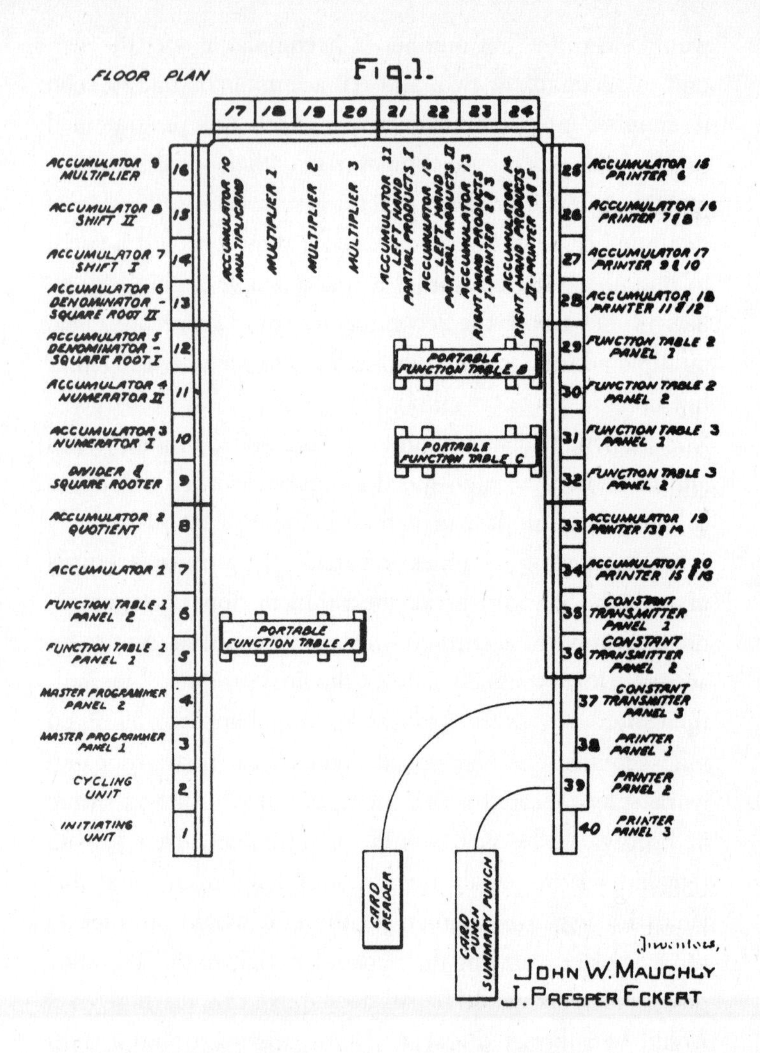

Floor Plan for ENIAC (The Hagley Museum and Library)

in the tens place, with the number 1 carried again. Finally, the machine would figure out that 6 could be subtracted from 18 three times. So the answer to the problem was 123. Finding the square root of a ten-digit number took all of 25 milliseconds!

Function tables—circuit panels that were on wheels and could be rolled around and plugged in at different points with a huge cable the size of a fire hose—were the equivalent of read-only memory in today's personal computers. Each had 728 knobs that could be set to different values and might be programmed for air density data for altitudes of zero to fifty feet or set up with constants or a sequence of numbers.

Five thousand times a second, the cycling unit would emit a fundamental pattern of signals—a series of pulses. One of those pulses marked the beginning of a cycle. Pulses would trigger certain actions, such as transmit its stored number at a particular point in the cycle. Every unit had a program control circuit that could recognize an input signal, which would tell it what to do. It was as if ENIAC were saying, "Hey you, take this number and do the following . . ." The circuit also told the unit when to do the operation and how to emit the output. All units operated in synchronization, starting at the same time, finishing at the same time. Everything was coordinated by the central programming pulse, the metronome of the machine. Essentially, this pulse was like Babbage's program control cards, only electronic.

Numbers were either loaded from IBM punch cards or stored in switches on function tables or formed in the process of computation and stored in accumulators. Only one ten-digit number was transmitted at a time on the digit tray. Every unit could listen in, but program control circuits told which unit to use that number.

Operators could schedule parallel operations, so different sets of accumulators could be working different parts of a problem at the same time. Then two or three different results from the parallel operations could be combined during the next cycle in the program. Operators could also use two accumulators together for one twenty-digit number, instead of two ten-digit numbers.

One person could run the machine, and the average running time before something went wrong was between five and six hours. Preparing a program for the machine took a month or two, and setting up the machine took a day or two, often involving setting some 3,000 switches. Debugging a program could take a week. Despite its size, ENIAC was a very personal computer. You could wander around it, watch it work, snuggle up to its warmth. Operators struggled with its fragilities and swore it suffered serious mood swings.

"The ENIAC was a son-of-a-bitch to program," recalled Jean Bartik, one of the original programmers. "The number of function switches was so limited, programs had to be broken up into rather small pieces."

Indeed, programming itself appears to have been

largely an afterthought. The system was jury-rigged from the start because of the army's changing specifications. It did not help matters that Eckert and Mauchly were hardware designers, not mathematicians or "software" people. Their job was to provide a machine capable of crunching numbers, with less regard for how those numbers would actually be crunched. It was a classic case of the technology getting ahead of the human interaction, a problem that continues today. Just because engineers can build something snappy doesn't guarantee it will be easy to use. Remember DOS? New versions of Microsoft Word have many, many bells and whistles dreamed up by hyperactive Redmond, Washington, software developers. But do we need "hyperlinks" and "letter wizards," or do those features only confuse and complicate? For ENIAC, the basic question left to its users was: How would it really work? Amazingly, that issue was secondary to getting the thing actually to work.

The Women Who Programmed ENIAC

Six women were chosen from among the several hundred human computers to work on ENIAC, making them the world's first computer programmers: Frances Bilas Spence, Jean Bartik, Ruth Lichterman Teitelbaum, Kathleen McNulty, Elizabeth Snyder Holberton, and Marlyn Wescoff Meltzer. This corps was rife with mathematical talent; they had been recruited from colleges across the country to help

Jean Bartik and Frances Bilas programming ENIAC (Used by permission of the University of Pennsylvania's School of Engineering and Applied Science, formerly known as the Moore School of Electrical Engineering)

the war effort by going to Philadelphia and working on ballistic firing tables. They tamed the complicated Differential Analyzer, and some taught math to other women to expand their ranks.

The ENIAC programmers were sent to Aberdeen to learn IBM tabulating equipment, and then, in a pattern repeated throughout the development of the computer, they were given little instruction on how to make the thing work—not even an incomprehensible manual. All they had were block diagrams and wiring schematics, and the chance to quiz engineers. With that they had to figure out how to program problems into it and get the correct solution.

"It was the most exciting work I ever did," said Jean Bartik, who, like several of the other female computer pioneers, went on to a distinguished career in programming. "I loved working with the engineers. I loved going to work. The more you use your head, the more you enjoy your work."

The presence of the women had, as might be expected, a major influence on the project. Professionally, they proved very astute at developing a system to use the machine and at troubleshooting its "bugs." Even though the women had significant responsibility for the success of the project, they continued to be treated as clerks, although they were in fact programmers.

Socially, many of them paired up with the men. Dating was common throughout the project, which made sense, since few, if any, people had time for a social life outside the Moore School. Three of the six chosen for ENIAC ended up marrying engineers on the project.

Brainerd Fights for Ownership of the Project

As the machine neared completion, tensions within the administration began to escalate. Grist Brainerd, in particular, realized that maybe the harebrained scheme he had been dragged into might actually turn into something pretty big. He began looking for ways to assert himself, though he had little understanding of how ENIAC worked or what the team was up to. His title, after all, was project director, making him the likely leader of the band. And he became determined to find ways to take the baton.

Grist Brainerd had been aptly named, given his background. He was a battler, a street fighter. His father had died when he was three years old, and his family was poor. He started out as a copyboy on the old Philadelphia *North American* newspaper to earn money, then worked his way through college as a reporter, covering night cops in the Tenderloin district. He had little regard for Mauchly and Eckert, both of whom had grown up far more privileged.

"Grist Brainerd is a person quite introspective," S. Reid Warren, a Moore School faculty administrator, said in a 1977 interview. "He tends to form opinions about people and then stick to those opinions, more or less permanently. He does not react well to 'screw ball' people. And accordingly he didn't react well to Eckert, but I think he admired him greatly. And I think he later felt that Eckert was really the responsible person for the development and that Mauchly was somebody off in the distance somewhere. Whereas from my point of view . . . it looked . . . as if they complemented each other beautifully."

Brainerd claimed that Dean Harold Pender had threatened to fire him if ENIAC failed, which seems out of character for the mild-mannered dean who had never shown much interest in the project. In any case, Brainerd made his move in late 1944 to gain ownership of the project.

A group called the Applied Mathematics Panel had learned of the project and asked the Moore School to present a report on computing machines to a scientific conference. The request came to Brainerd, who took it upon himself to write the report, even though he understood little of the science. The reason seemed clear: Whoever authored this first public report of the first computer would likely be credited with its parentage. The spoils often go to the author listed first on any scientific publication.

Eight months after Brainerd started working on the report, Eckert and Mauchly learned of his effort. They blew up and confronted Brainerd about his intentions. He refused to tell them his plans, other than to give them his outline. He wouldn't let them read a draft of his report.

Eckert and Mauchly wrote to Warren, the faculty administrator, complaining of Brainerd's "secrecy": "In view of the fact that the subject matter of this report is a device conceived by us and concerning that of which we have a more intimate knowledge than has anyone else, this reticence can hardly be taken as accidental."

Warren was a close friend of Brainerd, but he interceded on Eckert and Mauchly's behalf. He forced the three principals to work together on the report. (This was an internal Moore School spat, so Goldstine did not become involved.) Eckert and Mauchly hammered out the technical sections, which Brainerd had been unable to put together.

The friction did not end there. MIT sponsored a conference in October 1945, and Brainerd tried to block Mauchly from attending on behalf of the Moore School. Brainerd had decided he and Eckert would attend, but Mauchly, protesting again, was eventually included in the delegation.

ENIAC's Belated—but Impressive—Debut

Despite the Allied success, the war was still far from over in the fall of 1944. Having found a technology that could extend his reach, Hitler began to drop V-2 rockets on London in September. And in December, he mounted a massive counteroffensive, culminating in the Battle of the Bulge. U.S. tanks and bombers turned back the huge German force, and Allied air raids on Germany were relentless. British and U.S. airborne troops landed in Holland, and the Allies won back Athens. Finally, by early 1945, it seemed the war in Europe might be nearing an end. The Big Three—President Franklin Roosevelt, Prime Minister Winston Churchill, and Generalissimo Joseph Stalin—met in Yalta in February 1945 to plan a postwar settlement. Two months later, in April, FDR died, followed eighteen days later by Hitler's suicide. Germany surrendered in May, and, after atomic bombs were dropped on Hiroshima and Nagasaki, Japan surrendered in August.

By the fall of 1945, just as the war had ended, ENIAC was finally ready to go to work. The hardware was working; the programmers had learned how to run it. The army received the good word. It had taken 200,000 man-hours of work and cost $486,804.22, but ENIAC was finished—too late for the purpose for which it was built.

What the army got was a thirty-ton monster that filled 1,800 square feet—the size of a three-bedroom apartment in some cities. It had forty different units, including its twenty accumulators, arranged in the shape of a U, sixteen on each side, and eight in the middle, all connected by a ganglion of heavy black cable as thick as fire hose. It was 1,000 times faster than any numerical calculator, 500 times faster than any existing computing machine. It could perform 5,000 addition cycles a second and do the work of 50,000 people working by hand. In thirty seconds, ENIAC could calculate a single trajectory, something that would take twenty hours with a desk calculator or fifteen minutes on the Differential Analyzer. (Today, a supercomputer can perform the same work in three microseconds.)

ENIAC required 174 kilowatts of power to run, about as much electricity as it took to run a large broadcasting station. It cost $650 per hour in electricity when the computer was *not* operating, since the filaments in the tubes had to be heated and the fans had to keep operating. The final

Left to right: Homer Spence, Pres Eckert, John Mauchly, Jean Bartik, Captain Herman Goldstine, Ruth Lichterman (Used by permission of the University of Pennsylvania's School of Engineering and Applied Science, formerly known as the Moore School of Electrical Engineering)

tally was 17,468 vacuum tubes, 500,000 soldered joints, 70,000 resistors, and 10,000 capacitors—circuitry which today can be reduced to an integrated circuit that fits on a lapel pin. ENIAC wasn't the electronic progeny of Bush's Differential Analyzer or Babbage's Analytical Engine or any of the other machines that came before it. At its core, it was a bunch of Pascalines wired together.

But ENIAC had intelligence. It had the ability to react to data; it was programmable. And for the first time ever, intelligence was all-electronic. Electricity could be used to "think."

With the war over and the need for firing tables evaporating, the army wanted to test its new toy by giving it something bigger to think about. The army decided ENIAC could be of use on a problem that was still pressing: nuclear weapons. For its initial testing, Manhattan Project physicists Nicholas Metropolis and Stanley Frankel were brought in to solve a problem on ENIAC that remains classified even today. It is known that it related to the feasibility of building a hydrogen bomb. The Los Alamos enclave had built the fission bombs that had been dropped on Japan, and now was pondering the hydrogen bomb (which would become the basis of today's thermonuclear weapons). Developed by theorist Edward Teller, who like John von Neumann had emerged from the Hungarian Jewish middle class to achieve scientific excellence, the hydrogen bomb would use the energy of a fission bomb explosion to heat a specified amount of deuterium and tritium (two hydrogen isotopes) and create a thermonuclear reaction. But instead of splitting atoms, the hydrogen bomb would fuse them, releasing many times more energy. To get the recipe just right, the Los Alamos scientists had to calculate what was happening inside the reaction at increments of one ten-millionths of a second. They had made some crude calculations with slide rules and educated estimates, and mathematician Stanislaw Ulam had begun to raise doubts about Teller's design. ENIAC correctly showed that Teller's scheme would not work, but the results led Teller and Ulam to come up with another design together. For a shakedown cruise, the Los Alamos problem was quite a challenge.

Metropolis, who went on to direct Los Alamos's computing efforts for many years, remembered those days as a time of wonderment. "We would talk about this earlier history of the evolution of the ENIAC as well as what the future was possibly holding. It was a marvelous opportunity for an indoctrination [to computing]. It was absolutely priceless, almost," he said in a 1987 interview.

The Los Alamos visitors also helped raise the security concerns around ENIAC. Even though the room that housed the computer had always remained locked and the project was classified, new precautions were taken. No papers were allowed to be left lying around the room, and Metropolis and Frankel were required to keep Los Alamos papers with them in a briefcase at all times. One day Eckert went to a nearby drugstore called Dirty Drug with Metropolis, who left the briefcase with all the papers there. The pair went racing back, and the clerk behind the counter produced the bag. "If it had had something of value in it," he told them, "you guys would have lost it."

By early 1946, the army and the University of Pennsylvania were ready to unveil their creation to the public. A committee was formed, and elaborate planning was undertaken for a dedication ceremony and press conference on February 14, 1946, which would be sure to spark headlines around the world. A demonstration was planned with a so-

Left to right: Pres Eckert, John Grist Brainerd, Sam Feltman, Captain Herman Goldstine, John Mauchly, Dean Harold Pender, General G. M. Barnes, Colonel Paul N. Gillon (The Hagley Museum and Library)

phisticated problem. Since they couldn't use the Los Alamos work, the programmers were saddled with running a ballistic trajectory on ENIAC.

Once loaded on the machine, however, the trajectory program wouldn't work; it had a bug. The women worked day and night trying to pinpoint the problem, but all they came up with was frustration.

As the deadline drew near, stately Harold Pender, the dean of the Moore School, arrived one evening with a paper bag. "Go for it," he told the programmers and left. Inside the bag was a bottle of liquor, the sight of which caused the women to laugh hysterically.

Finally, with only a few days remaining, Elizabeth Holberton woke up in the middle of the night with the answer: One switch was off by one place.

Five-minute talks at the dedication ceremony were assigned to General Gladeon Barnes, who was head of research and development for army ordnance, along with Eckert, Goldstine, Mauchly, and Brainerd. Rehearsal was held January 30, 1946. No expense was spared; the menu at the five-course dedication dinner included lobster bisque, filet mignon, and broiled salmon steak. The keynote speaker was Frank Jewett, the president of the National Academy of Science, who had played an important role in sparking the Manhattan Project.

The War Department press release credited Eckert with engineering and design, Mauchly with fundamental ideas and physics, and Goldstine with mathematics and technical liaison. Mauchly was thirty-eight years old at the time; Goldstine was thirty-two; and Eckert, labeled the "spark plug" of the group, was "only twenty-six."

Clearly, they knew they were onto something big. "As might be expected, the designers and builders of the ENIAC regard this machine as but a first step—although a momentously big one—in applying electronics to the numerical work of science and engineering," the War Department press release stated. The group even predicted smaller, faster, more flexible devices. "It is worthy of note that the ENIAC established the fact that the basic principles of electronic computing are sound. It is inevitable that future machines will be improved through the experience gained on this first one, and that they will find application in all the diverse fields where the tremendous volume of computations required seriously impede progress."

Mauchly, always the dreamer, tried to press the importance of the invention on reporters, telling them that the computer would someday make the price of a loaf of bread a little lower or might reduce taxes just a bit. "All kinds of things entered our minds about the effects of the computer," he recalled many years later.

Reporters were impressed. "The War Department tonight unveiled the world's fastest calculating machine and said the robot possibly opened the mathematical way to better living for every man," wrote the Associated Press.

Word quickly spread far afield, and the Russian government tried to order an ENIAC from the University of Pennsylvania within days of the public demonstration. The Russians were denied; ENIAC wasn't for sale.

A few months later, ENIAC was taken apart, carried through a hole knocked in the brick building at the Moore School, and trucked to Aberdeen, where it ran for eight years. It worked on hydrogen bomb problems, made predictions of Russian weather patterns so nuclear fallout could be mapped in case of war, aided wind tunnel design, and, of course, crunched ballistic tables. ENIAC was used in the early days of the missile program, and it helped design special devices like nuclear artillery shells. "We did a lot of calculations we never knew what they were for," said Joseph Chernow, who was chief engineer on the ENIAC at Aberdeen from 1951 to 1953. The machine was finally retired in 1955.

The computer age was off to a blazing start. ENIAC quickly showed it was not just a demonstration project but a useful, long-lived tool, a true workhorse. Even more than that, it was instant inspiration for the world of computing. As ENIAC captured the dreams of Eckert and Mauchly, it sent many others off to chase new dreams of their own.

CHAPTER SIX

Whose Machine Was It, Anyway?

Even before ENIAC was formally unveiled, it had begun to capture the attention of the scientific world. But that happened by chance, not by plan.

At the railroad platform in Aberdeen in the summer of 1944, with ENIAC still more than a year from completion, Goldstine had a fateful meeting. John von Neumann, the world-famous mathematician, happened to be waiting for the same train north. As a mathematician who had gone to several of von Neumann's lectures, Goldstine recognized him instantly. "I had never met him," Goldstine said. "I was an egotist, so I decided I'd go talk to this famous guy."

Von Neumann had been part of the scientific exodus to the United States as Hitler's stranglehold on Europe progressed. Between 1933 and 1941, 326 scientists and scholars emigrated across the Atlantic, with some of the best congregating at Princeton University in New Jersey: Niels Bohr, Albert Einstein, Stanislaw Ulam, Eugene Wigner, and John von Neumann. Many of the top minds, including von Neumann, became key figures in the building of the atomic bomb.

Goldstine recalled that von Neumann was polite but totally uninterested in him until he began talking about the exciting program he was working on—a machine that would be able to do three hundred calculations per second. "Suddenly he changed," Goldstine said. "He found the thing he'd been looking for." Curiously, even though the ENIAC project was classified, and Goldstine later admitted he had no idea von Neumann was involved in an even more secret war project at Los Alamos, he didn't hesitate to talk about the project to an outsider like von Neumann because of the mathematician's stature. It was equally curious that von Neumann had not known about ENIAC. As Los Alamos's representative, von Neumann had inquired at the army about computing machines and was told about Aiken's Mark I at Harvard as well as Stibitz's work at Bell Labs. But no one in the army thought enough of ENIAC to even mention its development to von Neumann. Now, having learned of the machine from a chance meeting at the train station, von Neumann soon took Goldstine up on his invitation to come to Philadelphia and see ENIAC under construction.

Eckert said of von Neumann, "I was not familiar with

John von Neumann (Used by permission of the Archives, Historical Studies-Social Science Library, The Institute for Advanced Study)

great mathematicians, so I hadn't heard of him. Von Neumann didn't mean any more to me than Joe Apple or something. I know Goldstine was very impressed."

At their first meeting in the fall of 1944, Eckert tested von Neumann by posing a particularly challenging problem to him, and von Neumann quickly solved it. "He grasped what we were doing quite quickly," Eckert recalled. Von Neumann began visiting regularly. He became a powerful influence over the project, stimulating discussion about computing theory and practice, and pushing forward ideas for the second machine.

ENIAC, by its creators' own admission, had been designed in a hurry, using standardized parts wherever possible. Different circuits were designed in less than six months by different groups working independently, and since the goal was getting them all to function together, none had the most efficient design. ENIAC had three hundred different program controls. "We constructed a machine which we later felt—even felt to some extent at the time—was considerably more complicated than it should have to be, not only in its logical structure but also in the actual physical structure," Eckert said in a lecture he gave in 1946.

The riddles of ENIAC had already been solved by the time von Neumann happened upon the Moore School. But, as the army hinted at ENIAC's unveiling, design of the next computer was already in full swing, and that's where von Neumann focused his attention in Philadelphia.

Designing the Second Computer

Eckert had started on his drawings for the second machine in January 1944, roughly ten months before von Neumann arrived, but many discoveries remained to be made.

In contrast to ENIAC, the next computer would be of a simpler, more elegant design, with the memory storage Eckert had already drawn up. It would have an internal memory for 2,000 ten-digit numbers, compared to 20 ten-digit numbers in ENIAC. And it would have just one-tenth as much equipment as ENIAC. Instead of the multiple circuits for moving numbers around, the new machine would be built on a single channel, making it easier to program. The miniaturization of computers, where more power is continuously packed into a smaller package, had begun, and von Neumann, the unexpected interloper, played a key role in its design.

Throughout their work together, Eckert held his own against von Neumann and won his respect. Mauchly and Eckert were working on a new adding circuit for the second computer, for example, and found it took ten tubes for one circuit. Von Neumann, employing some abstract logical symbols, cheerily announced at one meeting that the adder could be built with five tubes. Eckert said no. Von Neumann went to the blackboard and drew his five-tube circuit. "Well, that won't work," Eckert declared. One tube didn't have enough power to drive the next tube in the time allowed in von Neumann's design. Von Neumann finally acceded, sort of. "You are right," he said. "It does take ten tubes to add—five tubes for logic and five tubes for electronics."

Late in 1945, the army agreed to a supplemental contract for the second machine, to be called the Electronic Discrete Variable Calculator, or EDVAC. Project PY, as it was called, received initial funding of $105,600.

The contract, however, only exacerbated tensions that were developing around the project.

Scientists Versus Engineers

In retrospect, it seems as though pettiness had been included in the blueprints for the project. There was the tension between the Moore School administrator, Grist Brainerd, and Eckert and Mauchly. There was a budding division within the computer team itself, with von Neumann leading a faction of "big thinkers"—logicians and Scientists with a capital *S*—while Eckert was seen as the commander of the lowly technical corps, the engineers, with a lowercase *e*. Mauchly was caught somewhere in between, neither able to hold his own with von Neumann nor viewed as a great engineer, as Eckert was.

The ENIAC project team had taken on the attributes of a vacuum tube. Ideas bounced back and forth between the poles of scientists and engineers, each trying to be dominant and each thinking the other inferior. Yet nobody was clearly in charge of the project. Sure, Eckert, barely even a graduate student, was taskmaster of the engineers, and von Neumann had taken charge of the logical design of the second computer, even though he was only a visitor. Goldstine had become von Neumann's proxy, and Mauchly, who still had a teaching load to carry, had largely become subordinate to Eckert.

By the time von Neumann had established his presence, Eckert, like Mauchly, was beginning to feel like an outsider within his own project. Eckert and Mauchly had an image problem, in part because they actually did much of the labor—the wiring, the testing, the building, the fixing—while high-level scientists stood watch. The pair, then, were naturally viewed more as tinkerers than as thinkers. This issue still runs through institutions today and in fact is the same issue that prompted Mauchly to avoid becoming an engineer in the first place, despite his father's pressure in that direction. Engineering was cookbook stuff. The real genius lay in the logic of the machine—its design. Mauchly, who preferred to be called a scientist rather than an engineer, made the point later that he and Eckert were doing logical design work from the start; it was just overlooked when the big shots came around.

In his book about computing history, Goldstine, who became von Neumann's disciple, expressed little admiration for Mauchly's mind but strong respect for Eckert's engineering. Both, in Goldstine's view, were technologists.

"There was this kind of dichotomy where Eckert and I were technicians while the other people were logicians and Supermen of one sort or another. And so we were not really in on all the upper level politics and strategy," Mauchly said in a 1973 interview at the Smithsonian Institution. "It became a tension as time went on when we realized that there was not only a kind of explicit division of labor but it became an even more explicit condition of not only who was responsible for what, but who was capable of what."

The result of the internal wrangling was that Eckert and Mauchly were losing control of their invention. ENIAC was miscast as the creation of a huge scientific melting pot of talent, not as the work of Eckert and Mauchly. And EDVAC, the second computer, was well on its way to becoming von Neumann's creation, even though many of the improvements planned for that machine were mapped out before von Neumann arrived at the Moore School.

Biographers such as Norman Macrae have noted that von Neumann was especially adept at seizing other people's good ideas, articulating them, and advancing them. The resulting concept generally was more brilliant, more precise, than the original idea. Indeed, part of von Neumann's adopted role at the Moore School was to crystallize ideas into a coherent hypothesis and put elucidating labels on designs. In a 1976 interview, Mauchly recalled that as Eckert and others developed the memory of EDVAC in further detail, von Neumann began calling it a "hierarchy of memories." "The word 'hierarchy' was suggested by von Neumann," Mauchly said. "But the concept was one we suggested to him."

Von Neumann was a natural lecturer, and when he was at Penn, the project group would gather in a classroom and debate at a blackboard. He would pace, think, stare, and pronounce. He was an intimidating force. "I remember saying 'no' to him when everyone was saying 'yes' one day," Jean Bartik recalled of a later encounter, after the EDVAC design meetings. "Everybody glared at me as though I were stupid. Von Neumann, on the other hand, tilted his head to one side, thought a bit, laughed, agreed he was wrong, and went on to something else."

Eckert, curiously, was a lot like von Neumann in some regards. Both grasped ideas quickly. Both were impatient with others. Both thought by talking, and batted ideas and concepts back and forth like a ball inside a pinball machine. (Playing pinball was in fact Eckert's favorite hobby.)

"Von Neumann and [Edward] Teller [the eminent theorist who fathered the hydrogen bomb] had one characteristic in common," Eckert recalled in a 1980 interview. "If you tried to tell them something and they understood it, they'd cut you right off there, and not let you finish. Both of them would do this. More so than anybody I ever met. I do it to some extent."

Early in 1945, von Neumann decided to put down on paper the ideas for the EDVAC. He had been summoned to Los Alamos for an extended period while final work was completed on the implosion plutonium bomb, whose design had been refined by his own calculations. (The implosion bomb, dubbed "Fat Man," used conventional explosives to compress a plutonium core.) The Manhattan Project was working feverishly to prepare for the Trinity test—the detonation of an implosion device in the New Mexico desert. As a result, von Neumann spent time writing out the design and structure of EDVAC while in Los

Alamos, and he sent this work off to Goldstine to be compiled and typed up.

The First Draft: *A Critical Document*

On June 30, 1945, Goldstine's typewriter gave birth to a 101-page report titled *First Draft of a Report on the EDVAC, by John von Neumann.* It was a masterful piece of work that poetically equated the structure of the computer with the structure of the human brain and talked of circuits as "neurons." (Eckert thought von Neumann had oversimplified the engineering.) The *First Draft* was more than just a conceptual dissertation, however; it was in essence a blueprint for constructing a computer that could store a program in its internal memory and run calculations at very fast speeds, even though it was short on engineering specifics. Von Neumann was listed as the sole author because the report was only a first draft, according to Goldstine.

Eckert and Mauchly didn't pay much attention to the authorship since the document was intended to be an internal summary of their work. It had been written by von Neumann, and it did lend logical clarity to the effort of building the EDVAC computer.

"It appeared to be purely a characteristic of Goldstine and no one else that in producing this mimeographed version, Goldstine put no one's name on it except von Neumann's," Mauchly later recalled.

Indeed, Warren, the faculty administrator, wrote in a 1947 memo that "[Goldstine] asked if this material could be mimeographed in the Moore School for use solely by members of the PY staff and Dr. von Neumann. . . . As I recall, I asked Dr. Goldstine whether or not the material should be classified, and he stated that since it was for use only within the group working on the EDVAC and since it was not considered to be a formal report, no classification was necessary."

Von Neumann credited only one suggestion to Mauchly and made no mention of anyone else on the project. He did credit certain ideas to academics outside the project, such as Harvard's Howard Aiken. "It was damning with faint praise," Mauchly said. "Who was supposed to be the inventor of all the other ideas?" Other than those minor references, there was no attempt at specifying who had come up with different ideas, even those central to the computer design, such as the stored-program concept.

Eckert had set forth this concept in a three-page memo dated February 1944 and signed by Mauchly—well before von Neumann showed up. Eckert had described a system that could store electrical pulses, along the lines of the unit he had designed for his radar project, which predated ENIAC. The device he envisioned would replace much of the difficult mechanical programming the women were doing on ENIAC. Instead of plugging in cords and turning switches to program the machine, instructions would be stored electronically in the memory. The device would be much faster, and much easier, and it was a key breakthrough in computer design. When von Neumann began working with the group about September 1944, Eckert detailed his radar-device idea to the Princeton mathematician. Yet to readers of von Neumann's document, it all appeared to be von Neumann's work.

What happened next undoubtedly changed the course of the development of the computer and sparked a major intellectual feud that continues to this day.

Even though Goldstine had declined to classify the report because it was only an internal document, he sent twenty-four copies to academic colleagues of von Neumann, including some in England. Requests began coming in for more copies, and soon several hundred were in circulation.

Sensing they were losing control of even their own ideas, Eckert and Mauchly scrambled to produce their own report three months after Goldstine hit the mimeograph machine. Even though it was written before a contract had been signed with the army, Eckert and Mauchly described their draft as a "progress report on EDVAC."

The report, not nearly as eloquent as von Neumann's, took pains to stake out what Eckert and Mauchly thought were their ideas. For example: "The invention of the acoustic delay line memory device by Eckert and Mauchly early in 1944 provided a way of obtaining large high-speed storage capacity with comparatively little equipment. An automatic electronic calculating machine was then planned, using the delay line device for 'internal memory.' "

The report did credit von Neumann where the authors thought appropriate. "[Von Neumann] has fortunately been available for consultation. He has contributed to many discussions on the logical control of the EDVAC, [and] has proposed certain instruction codes for specific problems. Dr. von Neumann has also written a preliminary report in which most of the results of earlier discussions are summarized. In his report, the physical structures and devices proposed by Eckert and Mauchly are replaced by idealized elements to avoid raising engineering problems which might distract attention from the logical considerations under discussion."

The report received little notice. Unlike von Neumann's outline, Eckert and Mauchly's report was stamped CONFIDENTIAL, meaning it would not be distributed. The officer in charge of security classification? Goldstine.

"A very great influence in all this is the fact that John and I were under military classification and couldn't publish," Eckert said in a 1980 interview.

That was true, but Eckert and Mauchly often found themselves upstaged by others because throughout their careers, they seldom wrote about their work. "They didn't have the patience to do it," Warren said.

Goldstine later admitted that liberties had been taken in the *First Draft.* He credited Eckert with the "raw idea" of using the delay line from radar as a storage device but said von Neumann perfected it. "Not everything in there is his [von Neumann's], but the crucial parts are," he maintained. Von Neumann never admitted as much.

Many say he never claimed to be the father of the stored-program concept, and thus the originator of computing architecture, but he didn't disabuse people of the notion that it all stemmed from him. Von Neumann died in 1957, a scientific hero for his many accomplishments, not only in computing but as a trusted adviser to presidents and generals. Despite the controversy, he never conceded publicly that the ideas in the *First Draft* were not entirely his.

In the end, distribution of the *First Draft* document resulted in exactly the kind of frenzy that von Neumann and Goldstine had apparently wanted. Von Neumann believed that the greater good of science was best served by the widest possible dissemination of ideas. Academics, too, saw his making a blueprint for electronic computers available to universities as a noble and proper strategy, designed to advance science and prevent the new field from being monopolized by commercial interests.

The strategy worked. Machines were built at universities around the world based on that 101-page blueprint. Because his name was the only one on the document, the concept of a stored program was attributed to von Neumann. Often, the creators of machines thanked von Neumann and cited the *First Draft* report; press articles frequently mentioned that the new creations were based on the work of the great John von Neumann. Von Neumann lectured extensively about computers, and even Mauchly said his advocacy gave "immense propulsion to the idea" of automatic computing by stimulating scholarly pursuit and making it easier for institutions to raise research money. Eventually, the design of a computer—a design that largely remains in use today—became known as the "von Neumann architecture." Von Neumann in later years made many contributions to computer science, and he himself was often tagged as both the originator of the stored-program concept and nothing less than the father of the computer. Deserved or not, that moniker clearly traces back to the *First Draft*, which Goldstine, among others, claimed was "the most important document ever written on computing and computers."

"While the placing of the EDVAC report in the public domain was very satisfying to both von Neumann and me, it ended our close relations with Eckert and Mauchly," Goldstine said rather dispassionately.

Maurice Wilkes, a British computer pioneer who received a copy of the *First Draft* and used it to construct a machine called EDSAC, said von Neumann deserved the credit he received because he was so instrumental in promoting the science. Von Neumann, Wilkes argued, "appreciated at once the possibilities of what became known as logical design and the potentialities implicit in the stored program principle. That von Neumann should bring his great prestige and influence to bear was important, since the new ideas were too revolutionary for some."

But many of von Neumann's actions don't appear to be graced with nobility; they appear to be calculated moves to claim credit for the birth of the computer. On January 12, 1946, more than a month before the scheduled unveiling of ENIAC, the Moore School team read about von Neumann on the front page of the *New York Times*. He was credited with a proposal for an exciting new machine capable of electronic calculations. The story reported a collaboration between von Neumann and RCA's Dr. Zworykin, now more interested in computers than he was when he dismissed those naive, enthusiastic tube burners from Penn. Von Neumann had remained in close contact with Zworykin, and the two had indeed been talking about making their own machine.

A public relations official inside RCA had leaked word of the breathtaking new machine to the *Times*, which reported that it might be developed by von Neumann and Zworykin for the U.S. Weather Bureau. That was the ultimate indignity for Mauchly, whose main dream in life was to solve weather-prediction problems. Mauchly was so outraged he got on a train to New York and marched into the *Times* offices trying to find the source of the story, which quoted only unnamed officials. That trip led him to Washington, where he finally traced the story back to the RCA public relations department. As it turned out, his anxiety was overblown. No one picked up on the significance of the story, not even the *Times*. The story died that day.

Friends and family say Eckert and Mauchly never forgave von Neumann—or Goldstine—for the *First Draft* distribution. Eckert later called von Neumann an idea pirate. "He spoke with a forked tongue. He said one thing and did something else. He was not to be trusted," Eckert said. "He may not have [taken the ideas] deliberately in the beginning, but he certainly continued to do it deliberately. I never did anything mean to him; I don't know why he should do something mean to me."

In a 1991 speech delivered in Japan, Eckert showed he clearly remained bitter after more than four decades had passed. "In my opinion, we were clearly suckered by John von Neumann, who succeeded in some circles at getting my ideas called the 'von Neumann architecture,' " Eckert said.

Mauchly was a bit more sanguine but no less bitter. "He [von Neumann] took every credit offered to him," said Mauchly, who sometimes chalked it all up to the "Matthew effect." (That theory holds that discoveries are attributed to the person with the biggest reputation at the time.) And not only is EDVAC misattributed to von Neumann, a 1985 book on the history of computing even credited von Neumann with having been a key factor in the design of ENIAC!

Others at the Moore School agreed with Eckert and Mauchly but were timid about challenging von Neumann. S. Reid Warren, the Moore School professor who ended up supervising the EDVAC project, admitted there was a lack of courage. "From my point of view, this great genius whom I greatly admired double-crossed us in no uncertain way. Now as I say, I can't prove that. If he was so damn innocent that he did all this without thinking about the facts, then he's more naive than I thought he was," Warren said in a 1977 interview with computer historian Nancy Stern that is archived at the Charles Babbage Institute in Minneapolis. "Sometimes when I've had plenty of sleep and [am] lying awake nights I think of this and think it was my fault. But I don't know how I would have stopped it. . . . I don't think I would have had the courage to go to him and say, 'Look, why didn't you put Eckert's and Mauchly's names out front?' . . . I trusted him. If that's an error, it's an error."

Even today, former colleagues at the Moore School defend Eckert. "Far as I know, Eckert came up with the stored-program concept using a mercury delay line, and he developed that for moving-target indicators in radar. It was before von Neumann showed up," said Jack Davis. Added Brad Sheppard, a classmate of Eckert's who also was part of the team, "Von Neumann had very little impact on what we were doing."

There have been efforts to clarify the record. In 1980, an article in the *Annals of the History of Computing*, cowritten by von Neumann's friend at Los Alamos, Nicholas Metropolis, reported: "It is clear that the stored-program concept predates von Neumann's participation in the EDVAC design. That von Neumann is given credit for this fundamental concept is likely due to the fact that he wrote a preliminary report which summarized the earlier work on the EDVAC designs including the stored-program concept. Von Neumann contributed significantly to the *development* of this concept, but to credit him with its invention is an historical error." And in their 1996 book *Computer*, Martin Campbell-Kelly and William Aspray conclude that use of the term "von Neumann architecture" "has done an injustice to von Neumann's co-inventors."

Was von Neumann a public-spirited academic who appropriated the ideas within the Moore School for the good of society? Or was he simply a scientist with a big reputation who enjoyed the limelight and saw an important opportunity to further his own legend? Only von Neumann could have answered these questions definitively. But he chose to remain silent.

Some biographers have suggested that von Neumann simply picked up Eckert's ideas and raced ahead with them, making the contributions his and his alone. Yet many of the important pieces of the *First Draft* report were engineering structures, which were outside von Neumann's area of expertise. The notion that von Neumann figured out better ways to wire up devices to manage electrical pulses is hard to figure. John von Neumann did many great things in science. But taking credit for the ideas of others, whether by commission or omission, diminishes his stature.

Battles over Patents

Even though ENIAC was unveiled on Valentine's Day 1946, and EDVAC was mentioned at the press conference, Eckert and Mauchly still had not filed for a patent on the invention. Not that patents had been overlooked. The Moore School group had started working on the issue two years earlier, in 1944, and had held meetings in Washington on a patent as early as August 30, 1944.

The arrangement appeared simple enough. The army contract gave the contractor—the University of Pennsylvania—the right to file for a patent on the invention, but the government got an irrevocable, nonexclusive, royalty-free right to make, use, and sell computers for any purpose. Eckert and Mauchly had cut a deal with the Moore School giving them the right to file for the patent if something patentable resulted from the ENIAC project. As part of the deal, Penn and other educational institutions would be granted a license to build and use computers for noncommercial purposes. In addition, Eckert and Mauchly had already polled the project team asking if any of the engineers thought they had contributed inventions that they wanted to patent. There was nothing of substance, other than the overall system, which was Eckert and Mauchly's creation.

In early 1946, Colonel Paul N. Gillon wrote to the Moore School saying that the army's security classification would soon be lifted on ENIAC. "It is suggested that you make the proposed action known to any inventors who worked on the ENIAC, and who contemplate filing patent applications in connection with it," Gillon said. The letter was passed along to Mauchly.

But by that time, the Moore School was already regretting its indifference to the patent issue two years earlier when the president of the university had written to Eckert and Mauchly giving them carte blanche to file patents. The university now began arguing that the machine was developed under a wartime contract with the institution, and that Eckert and Mauchly were unfairly trying to commercialize a public project.

This standoff led to all kinds of problems. On the advice of their own attorneys, Eckert and Mauchly canceled a meeting with RCA in January 1946 at Princeton to talk about the EDVAC—just prior to the *New York Times* story that gave credit for the computer idea to von Neumann and Zworykin.

And what about Goldstine? Did he have a legitimate claim? He thought so and wanted his name on the patent application. Goldstine had worded the War Department press release to say the three developers of the ENIAC were

Eckert, Mauchly, and Goldstine. But Mauchly edited it and struck Goldstine's name, not wanting to jeopardize a patent application made by just the two. Goldstine's relationship with Eckert and Mauchly, which had remained cordial because the full extent of the damage from the *First Draft* still wasn't known, changed immediately. "Herman was a pretty vindictive sort of personality, generally," Eckert said in a 1980 interview.

In fact, the entire War Department press release had been a major wound. Mauchly's diary for Sunday, January 20, 1946, notes that in the first paragraph of the first draft, "the following were mentioned: Army Ordnance, Dr. von Neumann, the Institute for Advanced Study, Dr. Zworykin and RCA. Not until three pages later were there any references to the Moore School, Pres, myself, or Herman. This was obviously starting off on the wrong foot."

As attention was split between the ENIAC unveiling and the actual construction of EDVAC, a new issue arose. The university president's letter to Eckert and Mauchly had dealt only with ENIAC patents. Now, Irven Travis, a Penn professor called away to the war who returned to the school to become the new supervisor of research, was insisting that Eckert and Mauchly sign a university patent agreement giving Penn the patent rights to future machines. "All people who wish to continue as employees of the university must turn over their patents to the university," Travis stated at a meeting.

Such a policy was far more restrictive than past practice, even apart from ENIAC. Several of the faculty at the Moore School conducted research at the university and had companies on the side to market products resulting from that research. Many universities were lenient on patent issues, giving professors plenty of incentive to invent and innovate. But Travis, who had been in charge of the modest computing program at Penn before being called to military duty, and who felt he had missed out on all the excitement since Mauchly had taken his spot on the faculty, was pushing the issue with a hard-line stance.

Some thought the dispute was just a clash of personalities. Both Eckert and Mauchly were outsiders as far as the Moore School faculty was concerned. Mauchly was a journeyman at an institution known to hire Penn graduates for the faculty over graduates of other schools. And although some of Eckert's classmates at Penn, like Jack Davis, were quickly hired to the faculty upon graduation, administrators didn't quite know what to do with Eckert. Neither he nor Mauchly fit well into the genial confines of the faculty club. They were oddballs, even if they had invented the first computer.

"The sense was Eckert was so dominant, they couldn't control him," said Davis, who was still a graduate student at the time.

Penn adopted the stance that Eckert and Mauchly were simply greedy to be harboring commercial interests rather than being devoted to scientific ideals.

Eckert tried to put pressure on the university through

the army, writing to Colonel Gillon on March 21, 1946, to warn that he would have to leave the EDVAC project if he didn't get the same patent deal from Penn that he and Mauchly had on ENIAC. Penn was trying to impose new rules, he complained.

The next day, Dean Pender sent Eckert and Mauchly an ultimatum, a list of three demands. He insisted on an answer by 5 P.M. the same day. If they were to remain at the school, Eckert and Mauchly had to give up future patent considerations and "certify you will devote your efforts first to the interests of the University of Pennsylvania and will during the interval of your employment here subjugate your personal commercial interests to the interest of the university."

After receiving Pender's ultimatum, Eckert and Mauchly resigned, a mere five weeks after the glorious, regal unveiling of ENIAC. The pair maintained they had been fired by Travis, who basically agreed with the notion. Penn was working with Princeton on EDVAC, and some of the engineers under Eckert and Mauchly were appointed to fill in. Still, the project suffered huge delays and major problems. Maurice Wilkes's EDSAC machine, which was based on information in von Neumann's report, was actually finished first and laid claim to the title first computer with a "stored-program design" ever built in the world. And at Penn, EDVAC, when it finally was finished, was the last computer project the university undertook. In 1949, computer development there was finished.

In later years, according to Warren, who remained on the faculty, that patent policy would be looked upon as "very, very naive. We didn't go out of our way to help people, and our general attitude was, 'Let's make it so it's helpful to the human race and so on.' . . . So perhaps from the point of view of patents, they [Eckert and Mauchly] got a pretty bad spot."

Arthur Burks, a member of the ENIAC team, once noted that no one involved in the project had permanent status at the Moore School, and the university didn't move to establish a center or build a program. It was an enormous mistake. Penn might have become the early center of the computer industry. The school had a very important lead on MIT and Harvard, which were both wedded to analog rather than digital methods for years. Philadelphia could have been what Boston became—a technology center with a huge, highly skilled employment base. "Perhaps it would have made a difference if they [Eckert and Mauchly] stayed," said Ralph Showers, a faculty member who remained at the Moore School for many years. "It's quite true that the university might have been more forceful about pushing developments in this field."

The university, according to Goldstine, was stupid; if it had gotten its engineers to sign an agreement long before ENIAC was finished, the outcome would have been much different. "They [The Penn administrators] asked too late. They blew the whole thing," he said. "When they didn't have the war to hold it together, it all fell apart."

Much later in life, when Eckert learned that Mauchly had decided to donate his papers to Penn, he barked orders to his wife. "When I die, nothing goes to them." Eckert was still angry about the patent dispute that had occurred nearly fifty years earlier. Since Eckert's death in 1995, several institutions have been jockeying for his papers, including the University of Pennsylvania. But as of this writing, they sit in boxes in a suburban Philadelphia attic, along with many of the toys and projects of Pres Eckert's youth.

CHAPTER SEVEN

Out on Their Own

Leaving Penn actually had been on Mauchly and Eckert's minds for some time. Both saw the computer for what it was: a tool to do other things, not an end unto itself. Neither had the drive to make the computer a scholarly pursuit. From the start, both Eckert and Mauchly intuitively recognized that the computer had enormous commercial application, and they both thought its development would be far faster in the business world than in academe. Remarkably, they understood even in 1945 that the computer's true strength lay in what it could do for others in business as well as in government. After all, Mauchly's own interest had arisen from the need for a computer to help forecast the weather. It could do much more. That was one reason they were driven to build a general-purpose machine, which could handle accounting problems just as easily as differential equations.

Others, like von Neumann, saw the computer more as a new field of study to be investigated, a riddle to be solved, just as nuclear physics had been for his Los Alamos compatriots. At a time when many scientists became world-renowned heroes and digital computers were a brand-new area of scholarly pursuit, the field offered excitement and the possibility of academic acclaim. There were advances to pioneer, papers to publish, prizes to be won. Curiously, the administrators at the Moore School never saw things that way; like Eckert and Mauchly, they were more concerned with the computer's commercial applications and the potential riches it represented.

Though the resignations happened abruptly, Eckert and Mauchly had been laying the groundwork to go out on their own. For more than a year, dating back to the time only two accumulators were working on ENIAC, Mauchly had been visiting with the U.S. Census Bureau and the U.S. Weather Bureau during his consulting trips to Washington, exploring their interest in computing machines. And Eckert had been talking with his father about how to turn his invention into a business.

Eckert's father played a key role. Contemporaries say Eckert felt enormous pressure from his father to make a million dollars—to match his father's commercial success. John Eckert Sr. understood that the new machine offered an entrepreneurial opportunity, and he was impatient as always. He convinced Mauchly's wife, Mary, that the two inventors should set up their own company and make computers to sell to big businesses and government agencies. At their respective homes, both men were being pressured to set out on their own. The Eckerts knew the monied people in Philadelphia who might invest in a new company, the bankers who would lend the capital. Eckert's father knew how to make it happen.

Still, other opportunities were tantalizing Eckert and Mauchly. Von Neumann, seemingly oblivious to the hard feelings over the *First Draft* report, was trying to entice Eckert to take the position of chief engineer at Princeton's Institute of Advanced Study, which was about to plunge into computer research and development under von Neumann's direction. Von Neumann had already hired Goldstine. Mauchly thought he had an offer at Princeton, too, though Goldstine and others later said Mauchly was never extended an invitation to join the faculty. Von Neumann had little regard, and no need, for Mauchly. Von Neumann and Mauchly were blinded by antipathy toward each other.

When Eckert turned them down, Goldstine and von Neumann ridiculed him for what they saw as shortsighted, greedy inclinations, obviously fostered by his capitalist father. Eckert's response was an accurate prediction of developments in the computer field. "I said, 'Well, I happen to think that the use of computers will be more furthered by manufacturers who make them quickly and cheaply as possible so a lot of people can get their hands on them than it would be to perfect them somewhere in more detail for a long while in universities,' " Eckert recalled in a 1980 interview.

"The snobs in this matter were not me or John. The snobs were von Neumann and Goldstine. . . . They attempted to make it look as if we were, you know, a bunch of Shylocks; all we were looking for was money, we didn't really care about whether people got good computers or not, or did mathematics. That's nonsense. We knew the only way this was going to fly was to get a lot of commercial money behind it."

At the same time, IBM's Thomas Watson Sr., aware of ENIAC because of IBM's role in supplying the punch-card machines for input and output, was curious about the potential of computers, though certainly not sold on their importance. The man who was said to have uttered the famous line "The world will only need five or six computers" tried to hire Eckert and Mauchly in 1946, offering to set them up in a computer lab. Eckert was interested; Mauchly didn't trust IBM, which he thought gouged people who needed the punch-card machines. In the end, both decided they could do better financially on their own.

In an April 1946 personal letter to a friend, Mauchly wrote: "We felt that the time was at hand when commercial development of electronic computing machines was going to be done—if not by us, by someone else. Some big companies were after the Moore School to make some arrangements with them about computing machine work, and it looked as if staying on would mean we would be indirectly tied to commercial organizations with the school as a middle man. We decided that we would rather have a more direct influence on our destinies."

Whether because they knew they worked well together and complemented each other, or because both were afraid to go out on their own without the security of the other, Eckert and Mauchly made a decision to stick together, no matter what. Eckert did not really consider staying in school and earning a Ph.D., because he was impatient and because he recognized the potential of the computer and the importance of the moment. Mauchly didn't want to give up his tinkering. "We got together and we did this thing [ENIAC], and I don't think either of us would have done it by ourselves," Eckert said.

Launching a New Industry

As it is with any business, getting started seemed in many ways to be the most challenging task. Eckert and Mauchly were able to raise some money from friends, but it was a tough sell. Computers were an unknown, and investing in them seemed to be folly to many people. Did the company have a product? Not yet. Did it have customers? Not yet. What would these machines be used for? Oh, lots of amazing things, maybe. Would they be reliable and easy to use? Well . . .

There was major distraction as well. Eckert and Mauchly had signed a contract to help teach a special summer course in computing at the Moore School. It wasn't just any course; it was an effort to teach the top people in electronics and mathematics from the United States and Great Britain about electronic digital computers. The course, "The Theory and Techniques for Design of Digital Computers," ran from July 8 to August 31, 1946, and would be taught by a handpicked group of computing experts. It was an unusual, extraordinary effort to jump-start research in the field, orchestrated by the Pentagon through the navy's Office of Naval Research and the Ordnance Department.

Eckert gave eleven of the forty-eight lectures; Mauchly and Goldstine delivered six each. Arthur Burks made three presentations; von Neumann was scheduled to give one talk, but never made it. The remainder was spread among various invited academics and military officials. One of the main issues remained whether digital computing was truly better than analog computing; Harvard and MIT were still not convinced. "The Moore School Lectures," as the course became known, turned out to be much more important than its organizers could have ever hoped: The course was a vital episode in the advancement of computing. Its students returned to their institutions and got to work swiftly developing this new field.

A lot of work went into the lectures, and it was a pressure-packed atmosphere, considering all the bad blood in the Moore School. The strain took away from efforts at starting the company, adding to the uncertainty of the time. Mary Mauchly wrote to her mother expressing deep worry about "Johnny's" future and saying she wished "they'd hurry up on the Eckert-Mauchly Company."

But before actually starting a company, Eckert and Mauchly wanted to have a contract in hand. Negotiations were held with several agencies, but only two seemed interested in funding computers, and one, the Office of Naval Research, had already hitched its wagon to MIT and Harvard. That left the National Bureau of Standards, a unit of the U.S. Commerce Department devoted to research. It was the U.S. Census Bureau, also a branch of the Commerce Department, that wanted the computer, and the way to finance it was through the Bureau of Standards, which was designed to fund inventors. The agency was enthusiastic about the project, but before signing a contract, the Bureau of Standards had to vet the Eckert-Mauchly proposal through an "expert." It chose George Stibitz of Bell Labs, who was still more enthralled with his own analog machines.

"I find it difficult to say much about the Eckert-Mauchly proposal. There are so many things still undecided that I do not think a contract should be let for the whole job," he wrote. Stibitz suggested a small contract to study the idea.

The Bureau of Standards, however, ignored the advice and, based on Assistant Director John Curtiss's interest in the project, decided to go ahead with a full contract, taking a chance on the two young inventors. The contract for the machine, then called EDVAC II, took the form of a series of grants totaling $270,000 over two years. That was cheap—not nearly sufficient—but Eckert and Mauchly figured it was a start.

With a preliminary commitment from the agency in hand and papers being drawn to launch their company, Mauchly decided it was a good time for a quick escape, a vacation with his wife to relieve some of the stress of the past several years. All through the ENIAC project, she had been working for the army at Penn teaching math to the ballistic-table "computers" and working with the Differential Analyzer. Now there was time for a break, and there likely would be few chances for getaways once the company was up and running.

The couple left their two children with John's mother and ran off to Wildwood Crest at the New Jersey shore, not far from Philadelphia. Giddy with their freedom, trying to throw their cares out to sea, the Mauchlys even jumped into the surf nude at midnight.

"It was a foolish thing to do," Mauchly said later. "We never did anything like that before."

Though the surf wasn't high, Mary was knocked down in the water. She was only a short distance from John, and she screamed once. He tried to reach her, but the waves knocked him down twice. Neither was a strong swimmer, and she was swept away by the current. John, flailing about, lost his glasses in the water, and a fog seemed to come up "and it clouded her from view," he said later. He couldn't see her, couldn't hear her, couldn't find her. He ran naked to the nearest lighted house to get help, but it was too late.

Mary Mauchly, John's wife of sixteen years and the mother of their eleven-year-old son and seven-year-old daughter, drowned September 8, 1946. Her body washed ashore two hours later, about two blocks from where she had disappeared.

Newspaper accounts of the tragedy said Mauchly was subjected to nine hours of questioning by the Cape May county prosecutor. The drowning was declared an accident.

Mary Mauchly's death left John a single parent, but his mother was nearby to help with child care. The loss brought an outpouring of support from friends, and even a sympathy note from von Neumann, which Mauchly saved in his files.

Despite the tragedy, the contract with the Bureau of Standards progressed. It was officially signed September 25, two weeks after Mary was buried. The partnership was able to raise some money from friends, and Eckert's father co-signed a loan for $25,000. In October, Eckert and Mauchly officially formed a partnership and set up shop in downtown Philadelphia across the Schuylkill River from Penn. This time, there was no doubt about titles: Mauchly was president; Eckert was vice president.

They considered five different names for the partnership and chose Electronic Control Company. Rejected were: Electronic Calculator Company, Electronic Manufacturing Company, Automatic Electronic Control Company, and Automatic Electronic Company. None had the word "computer" in the name because it wasn't part of the accepted lexicon and might scare off potential investors and customers. It was too new, too unfamiliar. Besides, most people still associated the term "computer" with women who did calculations by hand. The original proposal simply said they would develop "electronic calculating devices." "Electronic" was the one word all the considered names had in common.

Never did they underestimate the potential. The original business plan talked of devices for scientific laboratories, universities, and research foundations; industrial research and engineering; government agencies; accounting and bookkeeping departments of large business firms; inventory, stock control, and planning departments of large companies, insurance agencies, and other institutions with voluminous files, including libraries. These new machines, they suggested, could be used to instantly register transactions made at numerous remote points, which would be useful to banks, department stores, stock exchanges, train operators, racetracks, and so on. Automatic high-speed machines could be used in navigation, communication, typewriters and printing devices, and even for improving television. They could control high-speed knitting machines and numerous other industrial processes. They even mentioned electronic musical instruments. In other words, Eckert and Mauchly laid out the computer revolution with incredible insight and accuracy in their 1946 business plan.

Curtiss at the Bureau of Standards recognized that the company he was betting on was undercapitalized, so he urged other government agencies to contract for Eckert-Mauchly computers. The company did win additional contracts from the Air Controller's Office, which administered airplane traffic, and the Army Map Service. Electronic Control Company seemed to be off to a good start.

Patent Issues Pending

However, relations with Penn only worsened. In March 1947, Penn gave a presentation on its EDVAC, a project left foundering in development because of the defections, and Eckert and Mauchly were never mentioned. Mauchly wrote to Dean Pender complaining about the exclusion. The next month, April, a meeting was held at the Moore School on the still-unresolved issue of EDVAC patents. ENIAC patents were Eckert and Mauchly's to file, though they hadn't yet completed that task. But who actually had claim to EDVAC patents was an unresolved issue; there had been no prior agreement, and EDVAC wasn't covered by the original letter from Penn's president to Eckert and Mauchly.

The meeting had been arranged by the army, and von Neumann and Goldstine both showed up. It turned out that the pair had already approached the army to make a claim on a patent on EDVAC. On March 22, 1946, more than a year earlier, von Neumann had met with the Pentagon legal department about the patent situation and had filed an Army War Patent Form himself. The supporting documentation he included was a copy of his *First Draft* report. But the army's patent office realized it would face conflicting claims from von Neumann and Goldstine on one side, and Eckert and Mauchly on the other, so it arranged the meeting at the Moore School.

Von Neumann and Goldstine, who so passionately advocated putting developments in the public domain for the benefit of all, showed up at the Moore School with their own patent lawyer. Dean Pender and Irven Travis, the administrator whose zeal for a new patent policy triggered the breakup of the EDVAC team, protested since the university didn't have any lawyers at the meeting; Eckert and Mauchly hadn't brought lawyers, either, though they certainly had previously hired legal help for their patent claims. Von Neumann asked his lawyer to leave. He did not stake a claim to the delay-line memory at the meeting and said he did not expect to file any patents covering "such joint work."

The meeting held a major surprise for all principals, however. Army officials said the *First Draft* report, so widely distributed, most likely constituted "prior publication" of the invention. It had been more than a year since the document was "published," so the material was considered in the public domain and thus unpatentable. A major lawsuit between the two factions was avoided, but at a very high cost. Both parties lost, though history credited authorship to von Neumann.

Amazingly, Eckert and Mauchly still had not filed the ENIAC patent, even though the machine had become something of a celebrity itself. The army was even using ENIAC in a 1940s version of the "Be All That You Can Be . . ." recruiting ads, showing a picture of the machine and claiming that soldiers had the chance to "get in on the ground floor of important jobs." Eckert and Mauchly had started working on the patent in 1944, and in August of that year, the pair had a meeting at the U.S. Patent Office in Washington to discuss a patent on a "computer." (Eckert and Mauchly visited Mauchly's old friend from Iowa, John V. Atanasoff, who had made his own early machine, after the trip to the Patent Office.) But the patent work had been overtaken by the rush to set up the company. Now the EDVAC debacle was impetus for a stepped-up push to patent ENIAC.

On June 26, 1947, three years after they started on the patent, they finally filed a two-hundred-page document, written by Eckert and a patent attorney. The application was broad and unfocused, and it attempted to make more than one hundred claims covering the computing waterfront. The two inventors assigned their patent rights to their company to make it easier for the company to raise money.

The First Year in Business

In August, with a staff in place, Eckert and Mauchly officially began working on their new computer, which had been renamed UNIVAC, the Universal Automatic Computer. The office was in a converted dance studio on Walnut Street between Twelfth and Thirteenth, three blocks from Pres's father's office. The company had three floors above a ground-level clothing store. The engineering area still had the dance studio's mirror and barre, but otherwise Eckert and Mauchly simply replicated what they had had at Penn. The company had a casual atmosphere but with strict security precautions. Employees worked in locked rooms, and visitors were discouraged. Instead of hanging out around the Penn campus, they began spending hours at Robert's Sandwich Shop, writing out ideas on napkins. "Everything all day long was ideas," said Isaac Auerbach, the seventh employee hired at Electronic Control Company, in a 1972 interview.

From the start, they worked long hours, six and seven days a week, just as they had on ENIAC. To relieve tension, Eckert's father sometimes took engineers deep-sea fishing off the New Jersey coast, and Eckert hosted managers' steak fries parties at his house. It was a collegial atmosphere, with "Thank God It's Friday" lunches, even though Saturday was a regular workday.

Eckert and Mauchly had hired their engineering staff from Penn—despite their promise not to raid the school—and from other institutions. One of the things Eckert paid close attention to in interviews was hobbies; people with technical hobbies often were hired, because he figured they would better tolerate working on technical projects around the clock without breaks.

Earl Masterson recalled being interviewed by Eckert and Frazier Welsh, Eckert's engineering sidekick, on a Saturday morning. The interview lasted seven hours. Masterson, who worked for RCA in nearby Camden, New Jersey, brought along a book of photographs of projects he had worked on at RCA. Eckert sat either cross-legged on top of the conference table or perched atop the back of a chair so precariously Masterson thought he would fall. He opened his book to the first page, and Eckert asked about the contrivance. As Masterson explained, Eckert and Welsh would take off extrapolating other uses. The discussion would go on for ten or fifteen minutes between the two of them, Masterson said, and then he would flip over to the next picture and a new extrapolation would begin. "It was such an amazing sight, I guess one of the reasons I joined the company is because I wanted to be sure that what I saw was true," Masterson recalled in a 1988 interview.

Eckert was even more tenacious than he had been on

the ENIAC project at Penn. In short order he was discussing wild ideas about ink-jet printers, random access files, the need for nonmechanical input/output devices, and technologies that might be used to make computers smaller. Eckert was convinced that computer memories would have to be faster and less expensive. Some of his ideas were fifty years ahead of their time. Most have been borne out by subsequent developments. "Eckert was always a believer that if you press the state of the art, you can eventually make it work," said Auerbach.

As was his custom, Eckert came in late, at about 11 A.M., and worked well into the night. As pressure built, Eckert grew even more demanding. Jack Davis remembered mornings when Eckert would come in and glance first at a particular circuit, not the person working on it. "The first thing he'd say was, 'That resister is the wrong value.' Not 'Good morning' or 'How are you?' He'd instantly pick up where he left off last night."

Delays were rampant, many of them the result of Eckert's never-ending tinkering. Every time something was designed, he had a change. To Eckert, there was always a better way. The company rarely communicated by memos. It was all talk. Eckert and Mauchly were both addicted to working out their ideas by talking with anyone who would listen, Brad Sheppard recalls. Others remember Eckert summoning a worker into a conference room and delivering a two-hour lecture on one of his ideas, never asking for any reaction or input. Sometimes he would simply bombard the night watchman with his thoughts. Routinely, Eckert would stop by the engineers' office at quitting time and start talking, spitting out ideas. When he ran out of ideas, he would get up, say, "Well, okay, see you tomorrow," and walk out the door. Staffers would look at each other, knowing that in thirty seconds he was going to come back with another idea.

Mauchly remained the laid-back, stabilizing force who concentrated on systems and programming—the logic of the computer. He and Elizabeth Snyder Holberton, one of the programmers who, like Jean Bartik, had joined the company, worked on various ways to give computers instructions.

Sometime within the first year after his wife's drowning, Mauchly started dating Kathleen McNulty, who went by the nickname Kay, another of the six original women programmers who had remained with ENIAC and was working with it in Aberdeen. McNulty's parents frowned upon the relationship because Mauchly was considerably older, was a widower, and wasn't Catholic. Besides, there was even a bit of mystery surrounding Mary Mauchly's drowning. But the two continued dating, visiting frequently. "Life is very lonely when you're not around," Kay wrote to John on September 9, 1947. "I have a good remedy for this, of course, and plan to take my own prescription." The two were married less than six months later; Eckert was the best man.

In some ways, it was already a golden era for computing. Mauchly later recalled that in the early days of UNIVAC development, the team would sometimes come up with one hundred inventions a day. "Nobody does that anymore," he lamented.

Grace Murray Hopper, the famed programmer who later developed the COBOL language, left Howard Aiken at Harvard to join Eckert and Mauchly in Philadelphia, working closely with Mauchly and Holberton to develop an instruction code for UNIVAC. She remembered it as a time when opportunities were open to women in the workplace, even in what later was to become a very male-dominated industry. "Eckert and Mauchly were singularly unprejudiced but also they were trying to gather a team to build that first computer which no one believed in. . . . It was a very flexible group. It was brand new. It had no traditions of this person had that job or anything else," Hopper said in a 1976 interview tape-recorded by the Charles Babbage Institute.

But there were many problems, too. Mauchly wouldn't really accept responsibility for managing, even though he was president of the company. He was more like a benevolent colleague than a president. One time an employee asked for a raise, and Mauchly agreed. Then the fellow went to the bookkeepers, who told him there wasn't enough money to give him a raise. When the employee went back to Mauchly, he just shrugged.

There wasn't much management at the company, period. Mauchly's side was a democracy; the engineering side was a dictatorship. The result was as effective as the United Nations—there was lots of hard work and little achievement.

In fairness to his management skills, Mauchly did understand the financial problems the company faced, however. He was often called upon to hustle up new investors, to meet the growing need for capital. He even took on consulting jobs, working for Northrop Aircraft Company, for example, as a computer consultant.

By October 1947, Electronic Control Company was so starved for cash it signed a contract for another type of computer, even though the UNIVAC project was still in its infancy. Eckert and Mauchly agreed to build a computer for Northrop that would control the flight of a new missile called the SNARK. The computer had to be incredibly accurate and unfailing, and Northrop even wanted it to be able to be transported and operated in an airplane—a rather preposterous idea at the time, given the size and fragility of the early machines and their voracious appetite for electricity. The design Eckert and Mauchly proposed was to put two small computers together and run them in tandem. The machine would be called BINAC (Binary Automatic Computer), and it would sell for the equally preposterous price of only $100,000. It would be smaller and a bit simpler than UNIVAC, so Eckert and Mauchly thought they could throw together BINAC and not get too far behind on UNIVAC.

To raise additional money, they decided to branch beyond the partnership and sell some stock. On December 22, 1947, they incorporated the firm as the Eckert-Mauchly Computer Corporation. It had thirty-six employees at the time.

Some of the internal problems were obvious even to Mauchly. On February 5, 1948, he wrote a memo to employees: "The more I think about the situation in which we find ourselves at present, the more I am convinced that we are losing a hell of a lot of valuable time by reason of the fact that we are slow in making some necessary decisions."

Eckert-Mauchly Computer Corporation (EMCC) found itself in a catch-22 common to start-up firms: It couldn't get orders until its finances improved, but it couldn't improve its financial position until it got orders. Capital was still hard to come by for computers, since most investors didn't see it as a developing market. The company didn't want to rush into stock deals that would force it to sell a big chunk of the company and lose control, but its position was very weak. "They lacked a salesman, a person with selling talents," said Brad Sheppard, a Moore School veteran who signed on with Eckert and Mauchly.

In fact, the salesmen they did have were technically inclined; therefore they understood the limitations of their machine. Instead of hyping the product, they actually dampened potential customers' expectations. In contrast, many successful technology companies have succeeded by driving expectations high, creating a "buzz" about a new product, and preannouncing capabilities, even if those innovations were years away. The trick is to make the new product sound so exciting you have to have it, and if you don't buy it, your competitor might get an advantage over you. But that wasn't EMCC's style. "From the beginning the UNIVAC I sales effort was insufficient, unaggressive and unimaginative," said Saul Rosen of Purdue University in a 1968 historical survey of electronic computing.

Negotiations were disastrous as well. The day the National Bureau of Standards came to Philadelphia to negotiate terms for the first UNIVAC, Auerbach says Mauchly walked into his office at 8:30 A.M. and handed him a copy of the proposal and said, "Would you come in and help negotiate this contract?"

The big mistake was that they negotiated fixed-price contracts, rather than cost-plus-developmental contracts. Despite not knowing what it would cost to invent these machines, Eckert and Mauchly sold them like completed products pulled off the shelf. It was a way to lessen the risk to the buyer and undoubtedly generated more contracts. But it was a fiasco for the company. When the money was gone and the machine wasn't completed, they would be in a terrible bind. BINAC, for example, cost the company $280,000 to produce, but Northrop paid $100,000. It took more than $900,000 to get the first UNIVAC operating; the contract promised only $270,000.

Margaret Fox, who inherited the UNIVAC contract at the National Bureau of Standards, found a meltdown in the making. The company was obviously running out of money, and UNIVAC was far from completion. "When I dug into the contracts and into the background of it, I was just fascinated to think anybody would be so stupid. . . . Eckert and Mauchly for taking a fixed-price contract—just unbelievable," she said.

Scrambling, Eckert and Mauchly got Northrop to pay 80 percent of the BINAC funds up front, giving them a bit of money for a move to a bigger facility in the spring of 1948. The company relocated to the seventh and eighth floors of a building at Broad and Spring Garden Streets in Philadelphia. The highlight of the neighborhood was an old barber nearby who cut hair for twenty-five cents. He had a young, blond wife who also cut hair, so engineers always went two at a time to be sure one would get his hair cut by the woman.

That summer the company was at a major crossroads. Some, backed by Mauchly, were lobbying for a plan to stick with BINAC for a time and take its design and modify it for other uses. Maybe universities would buy a stripped-down version for their own research? Maybe companies unwilling to pony up for a full UNIVAC would test the computing waters with a simple BINAC? It was a way to build some sales. Others—especially Eckert—were advocating a plunge into UNIVAC. BINAC was a temporary distraction of limited use; UNIVAC was gold.

On August 18, 1948, Mauchly wrote a memo urging the sale and rental of BINACs as a "good avenue for improvement of our financial condition." The memo was presented at a meeting, and Eckert flew into a rage. "Literally, chalk and blackboard erasers flew about the room to express his rage," Isaac Auerbach recalled.

Instead, Eckert proposed building something simple and mundane—spark plug testers for automotive repair shops—as a way to raise some quick cash while developing UNIVAC. He worked all weekend designing a tester, then dropped the idea. Still, the upshot was that Eckert exerted more control over the company, and Mauchly took more of a backseat.

The company pressed on with UNIVAC—the technology was more important than the payroll. From that point forward, there was no doubt that technical considerations took priority over business strategy at EMCC. "In some ways, his [Eckert's] dominance [as well as Mauchly's submission to it], while a distinct asset technologically, was a barrier to the success of their business," wrote computer historian Nancy Stern in her analysis of the Eckert-Mauchly Computer Company.

Eckert and Mauchly grew farther apart. One time, Mauchly was shocked to find out Eckert's father had backed a bank loan and Mauchly hadn't known about it. But they managed to keep their partnership together. The two still felt that they needed each other.

Business Prospects Get Worse

The odds were stacked against them in many ways. The U.S. State Department had declared that Eckert-Mauchly couldn't sell UNIVACs abroad, even though several foreign governments—friend and foe—had made serious inquiries. Running out of money, Mauchly sometimes had to put the weekly paychecks in the company safe because they couldn't be cashed—there wasn't enough money to make payroll. At times the firm, which had grown to about one hundred employees, took three months to settle expense reports. A. C. Nielsen Company, the television ratings service, offered to invest but wanted 40 percent of the company's stock. Eckert and Mauchly weren't yet willing to give up that much control. Instead, engineers were asked to put up $5,000 each to buy preferred stock to help keep the firm afloat. Northrop sent an executive to Philadelphia to try to help the company out of its morass. And Mauchly was personally so far in the hole his own insurance policy was unpaid on its renewal date. His agent wrote him saying he paid it out of his own pocket so the insurance wouldn't lapse.

The grueling pace and corporate poverty took a toll on the staff, and loyal engineers began departing for less overwhelming jobs. Auerbach left shortly after Eckert's UNIVAC push. Jack Davis, pressured by his wife, resigned after a memo came out with a new work schedule, requiring work seven days a week, twelve hours a day. "The intensity was just horrendous," Davis recalled.

Past disagreements continued to haunt them as well. Eckert-Mauchly competed for a contract for a computer for the Office of Naval Research, which had tried to build its own machine but failed. The competition for the contract came from Raytheon Manufacturing Company, which was insisting on a contract covering its costs plus a fixed fee. The decision was up to the National Research Council, which was chaired by John von Neumann. Howard Aiken, the Harvard developer still wedded to electromechanical instead of electronic methods, was also on the council. Aiken considered himself a fierce rival of Eckert and Mauchly.

The council picked Raytheon, noting in its recommendation that it was impossible to build a mercury memory that could operate a computer—the very issue at the heart of the *First Draft* dispute with von Neumann. The mercury memory was one thing that Eckert-Mauchly had already designed and built, but no one on the Research Council went to Philadelphia to see it. Why not? "A great deal of personal animosity," Auerbach recalled.

Bad as things were, there was yet another major problem confronting the young company. Anticommunism fervor was taking root in the United States, later to develop into McCarthyism. In 1948 at Eckert-Mauchly Computer Corporation, it took the form of an Army Intelligence Division investigation of the firm for its security clearance. After all, BINAC was a top secret project.

The army investigation found that five of the nine people with security clearances at EMCC had "subversive tendencies or connections." The five were Bob Shaw, the ultraliberal engineer who had been a supporter of Henry Wallace and the Progressive Party; Brad Sheppard; Albert Auerbach, another engineer (not related to Isaac Auerbach); Dorothy K. Shisler, Mauchly's secretary; and Mauchly himself.

According to Mauchly's Federal Bureau of Investigation dossier, army investigators tagged him as suspect for two reasons. First, he was a member of the Philadelphia branch of the American Association of Scientific Workers, an organization that the army said was formed by the Communist Party as a front to influence legislation on atomic energy. Indeed, Mauchly had signed a petition in which 980 scientists urged the president and Congress to adopt civilian control of atomic energy. Although this notion sent the army into orbit, it was later adopted by the country.

The second reason for Mauchly's disqualification was even more curious: "Mauchly's wife was mysteriously drowned while both were moonlight bathing in Wildwood, N.J.," the document said accusingly.

Army Intelligence asked the FBI to investigate further. It did, and on November 11, 1948, the FBI delivered a fifteen-page report clearing Mauchly of misconduct or disloyalty. The FBI concluded, as others had, that all Mauchly was guilty of was being "eccentric."

The army, which had banned EMCC from receiving classified documents and thus from getting military contracts, wasn't satisfied and stood firm on its disqualification. On January 31, 1950, the army's Philadelphia Ordnance District sent a letter to the company saying security clearance was denied for the firm and Mauchly and Shaw. The company lost out on key defense contracts that could have kept it vibrant.

Financial and Real Estate Deals

Eckert and Mauchly did find a temporary financial savior, however. In 1948, a Delaware racetrack owner had approached them about developing electronic versions of the "totalizators" that tallied racetrack bets and posted odds. Advanced versions of these machines ran on relays, much like Stibitz's computers, and the racetrack thought Eckert and Mauchly might do better with electronic circuits. Besides, a Baltimore, Maryland, company called American Totalizator Company had a monopoly on the business, and the Delaware track was looking for a way to break that monopoly.

Eckert and Mauchly weren't interested in working for the track, which they saw as a special-purpose use not suited to a general-purpose machine. But the pair's patent attorney, George Eltgroth, decided to pursue the totalizator option. Eltgroth had been at Bendix Corporation in Baltimore before joining EMCC, and he knew executives at

American Totalizator. Eltgroth brought the vice president of American Totalizator, Henry Straus, to Eckert-Mauchly. Instead of a special-purpose use, Straus was willing to simply back the company financially and leave control with Eckert and Mauchly. For $488,000 plus a $62,000 loan, American Totalizator would receive 40 percent of the stock in EMCC and would have four seats on the company's nine-person board. Straus was named chairman of EMCC.

After only a year at its Broad Street location, and despite the fondness for the barber's wife, Eckert-Mauchly Computer Corporation was on the move again. Eckert's father, after all, was a real estate whiz always on the lookout for a good deal. In the spring of 1949, the expanding company moved to north Philadelphia, taking over a loftlike, cavernous former knitting mill at 3747 Ridge Avenue, poetically located across from a cemetery and next door to a junkyard. There was something symbolic, too, about taking over a former knitting mill, which had moved to the South. The high-tech transformation of the Northeast, replacing low-level manufacturing with computing, had begun, as had the continuation of Babbage's knitting-loom influence on computing. EMCC had two floors: The upper story was engineering, with only three offices, for Eckert, Mauchly, and Eltgroth; the main floor was manufacturing, with a stockroom in the basement. The biggest drawback, however, was that the place had no air-conditioning. One employee kept a thermometer on his desk, and some summer days it read 102 degrees.

Completion of BINAC

In August 1949, BINAC was operational. In its test program, the machine ran for forty-three hours without any stops or errors. BINAC had a mercury delay-line memory that could hold 512 words of information. Each word consisted of 30 bits of data or information, plus a parity bit that provided odd/even parity checking on memory. It was enough for simple programs. It ran at a speed of four megahertz—pretty snappy for its day. It had a keyboard, a character printer, and a tape loader that read magnetic tape developed by EMCC.

Surprisingly, BINAC lacked some of the quality craftsmanship that had become Eckert's trademark. It had been created in a rush, of course, but construction still had taken nearly two years. The real reason for the inferior craftsmanship was that Eckert always considered BINAC a stepchild and put his focus on UNIVAC. BINAC had a problem with plug-in circuits: If the machine was jostled a bit, some of the circuits could come loose. Hardly the thing you want in an airborne computer, but that notion had been given up long before. All Northrop insisted on in the contract was that the machine *fit* through the cargo door of a large plane. Yet because of the circuit problem, everyone tiptoed around BINAC to keep from shaking the machine.

Except at the BINAC completion party. Engineers attached a loudspeaker to the machine's high-speed data bus, the pipeline that carried pulses, creating "music" as pulses coursed through it. They also programmed the machine to release a hard-boiled egg from its innards on command. Fittingly, BINAC laid a big egg. After it was delivered to Northrop in California, it never really worked. Eckert-Mauchly blamed Northrop; Northrop blamed Eckert-Mauchly. Northrop paid the remainder due under the contract and set BINAC aside in a warehouse. Rumors spread, further damaging the reputation of Eckert-Mauchly.

"The truth is that it was never given the chance to work after it left Philadelphia," Mauchly claimed in a 1978 letter. "Eckert and I were anxious to get on with the design and construction of the UNIVAC for Census, so we spent no time crying over what happened to BINAC."

The UNIVAC Project: From Setbacks to Success

By 1949, Eckert-Mauchly Computer Corporation had 134 employees and contracts for six UNIVAC systems totaling $1.2 million, according to Mauchly's annual president's report. Prudential Insurance Corporation and the A. C. Nielsen Company had both signed up, though at the bargain-basement price of only $150,000 per machine.

The company photo that year showed the balding Eckert in his usual white shirt and black tie, tight-lipped, sitting upright with his hands crossed, shoulders pinned in. But the rest of the managers were obviously more relaxed.

Only three of the twenty-five men were in ties. There were four women. Mauchly, in a short-sleeved work shirt, was his usual lanky self, leaning back and kind of spread out, smiling with his head cocked.

The main UNIVAC construction effort started in earnest in the summer of 1949. Then tragedy struck Eckert and Mauchly once again. On November 25, 1949, Henry Straus, the American Totalizator vice president who had become a friend of and business adviser to the company, was killed in a plane crash. Straus had played the von Neumann role, without the competitive element. He had visited often and made many suggestions. As he learned about computers, he taught Eckert and Mauchly about business. He had taken them under his wing, and he had gotten EMCC to fly.

Yet, with Straus gone, American Totalizator's board wasn't interested in computers anymore. The company wanted its money back. Still some $500,000 undercapitalized, Eckert and Mauchly suddenly had to raise another $438,000 to buy out American Totalizator.

Late in 1949, Eckert and Mauchly went to Thomas Watson Jr. at IBM in New York, offering a majority stake in EMCC to the company that by this time was running full steam ahead with its own computer development program. Watson, who had previously met only Eckert, recalled Mauchly as a "lanky character who dressed sloppily and liked to flout convention." Watson wrote in his 1990 memoirs, *Father and Son & Co.,* that Mauchly slumped in the couch and put his feet on the coffee table. (Mauchly's family questions this account, saying that as one who was always respectful of property, he never put his feet on a coffee table, even at home.) The interview was in the elder Watson's office in New York, conducted by both father and son. Lawyers had already told the Watsons that IBM couldn't buy Eckert-Mauchly because of antitrust concerns; UNIVAC was one of the few office-machine competitors IBM had. The Eckert-Mauchly offer was turned down.

On February 15, 1950, just four years after ENIAC's unveiling, Remington Rand, the typewriter maker eager to get into the computers field, agreed to repay the $438,000 owed to American Totalizator and buy the remaining 60 percent of the stock in EMCC for about $100,000. Remington Rand also agreed to give Eckert and Mauchly a free hand with what became known as the Eckert-Mauchly Computer Corporation division, although they had to report to Leslie Groves, the retired army general who had run the Manhattan Project. The pair would get 5 percent of profits in the division for eight years—provided there were any profits—as long as they were not competing against the company. Remington Rand also agreed to pay Eckert and Mauchly half the income from patents during the term of the agreement and a guaranteed yearly salary of $18,000 each. It wasn't much of an offer, but they had no other choice.

The dream of their own company was over, and they

knew it. Groves tried to put in financial controls and renegotiate contracts, but delays continued. Prudential and Nielsen both dropped out, had their money returned, and bought their first computers from IBM.

In March of 1951, the first UNIVAC finally became operational. It took data in through an ingenious magnetic tape system the company had pioneered. UNIVAC's magnetic tape was durable and reliable and could read 1 million decimal digits at a rate of 10,000 digits per second. It was far superior to punch cards from an engineering point of view, just not from a marketing point of view. UNIVAC used about 5,000 vacuum tubes, compared to nearly 18,000 in ENIAC, and it consumed only fifteen kilowatts of electricity.

"That was a pretty damn big achievement," Mauchly said in his 1978 videotaped recounting of computer development with his friend Esther Carr. "Just five years after we left the Moore School, we produced the first commercial computer and the tapes with it."

The Census Bureau decided to keep its UNIVAC at EMCC and have it operated there by the company under contract, avoiding the problems Northrop encountered when BINAC was transferred. Still, heat was constantly a problem in the summer because the building had no air-conditioning. One day it was so brutally hot that tar on the roof started dripping through the ceiling on the brand-new computer, prompting a mad scramble. The first two or three UNIVACs were air-cooled, exacerbating the heat problem. Engineers cut a hole in the wall on the first floor and pumped air up through the bottom of the computer and out through the wall of the second floor. On summer days when the air temperature was too high to cool the machine, workers laid dry ice in the ductwork to cool it down. By contrast, it got so cold in the building in winter because of the holes cut in walls that workers had to don overcoats and gloves while tending to UNIVAC. Later, the company developed a system with chilled water running through heat exchangers in the bottom of the machine, cooling the machine with continuously circulated air.

UNIVAC

Remington Rand tried all kinds of ways to sell UNIVAC computers, even resorting to gimmicks. The company sponsored the television program *What's My Line* with advertisements for its Remington electric razors. So it took Eckert to New York for an interview with the staff of the show, who decided that designing and building a computer was, in Eckert's words, "not of sufficient general interest to put a person who had done that on their program."

The machine's big chance came with the 1952 presidential election. A Remington Rand executive named Art Draper, who was in charge of the company's research labo-

Left to right: unidentified technician, Pres Eckert, and Walter Cronkite (Courtesy of UNISYS Corporation)

ratory, came up with the idea of using the computer to predict the results of the election based on early returns. Draper got CBS News interested in the idea, and Walter Cronkite decided to go along with the experiment in the interest of entertainment, rather than serious news.

Programmers took early returns in key districts in eight crucial states: New York, Pennsylvania, Massachusetts, Ohio, Illinois, Minnesota, Texas, and California—and loaded them onto computer tape. The results were compared with past voting patterns.

CBS had a fake computer control panel on its election-night set in New York, using blinking Christmas tree lights for effect. Correspondent Charles Collingswood was actually at Eckert-Mauchly Computer Company in Philadelphia. EMCC had even enlarged the typewriter head used for the printout so the camera could broadcast the actual output from the machine.

IT'S AWFULLY EARLY, BUT I'LL GO OUT ON A LIMB
UNIVAC PREDICTS—WITH 3,398,745 VOTES IN—

	STEVENSON	EISENHOWER
STATES	5	43
ELECTORAL	93	438
POPULAR	18,986,436	32,915,049

A landslide for Dwight Eisenhower? All the polls had predicted a close race! The printout even listed the odds for a Stevenson victory at "00 to 1"—greater than 99 to 1, the highest odds that programmers had anticipated.

Both CBS and Remington Rand decided on the spot that they couldn't let that first report out; the risks of embarrassment were too great. Programmers went back and fiddled with parameters in the machine's calculations to make it a closer race, with Republican Eisenhower still winning. The prediction used on the air had UNIVAC giving the odds at 8 to 7 in favor of Ike.

In the final tally, Eisenhower won 442 electoral votes,

and his Democratic opponent, Adlai Stevenson, had just 89. UNIVAC's original prediction was off by only 4 electoral votes out of 531—better than 98 percent accuracy.

That night on the air, CBS confessed its disbelief of the computer and explained how the results had been fudged. (The problem with computers, it has been noted many times, is often people.)

UNIVAC became an instant celebrity, and sales improved. Remington Rand advertised it as the "first so-called 'Giant Brain' on the market." General Electric became the first nongovernment purchaser in the spring of 1953, buying machine No. 8. Then Metropolitan Life signed up, and U.S. Steel, E. I. DuPont, and Franklin Life. There could have been more if Remington Rand had agreed to let companies lease the machines, instead of insisting on outright purchases.

A total of forty-six UNIVAC systems were produced, and the last one wasn't deactivated until 1969—twenty years after it was designed.

Personal and Business Losses

Despite the apparent success, Remington Rand was a house divided. The company had purchased William Norris's Engineering Research Associates in St. Paul, Minnesota, in 1952 to complement its Philadelphia computer division. Instead, St. Paul and Philadelphia fought constantly, and the civil war undermined Remington Rand's efforts. More specifically, Eckert and Norris fought constantly. (Norris could stand only so much; five years later, he took the best people from ERA, including a man named Seymour Cray who would later become the father of "supercomputers," and founded Control Data Corporation.)

The stress took a huge toll on Eckert—and on his marriage. He was so preoccupied with work that his relationship with his wife, Hester, was often an afterthought. Jack Davis recalled going out to dinner with Pres, Hester, and Frazier Welsh, another engineer. Eckert and Welsh spent the entire evening working out details of problems, while Davis talked to Hester. After dinner, Eckert went back to work. Another time, Earl Masterson remembered accompanying Pres and Hester on a shopping spree. She wanted him to help pick out yard furniture, so his solution was to take along Masterson and discuss high-speed printer design the entire time. Often Hester came to the office in the evening with food and thermoses of hot coffee. Otherwise, Pres would not even stop to eat. While he ate, she stood idly by.

In 1953, troubled by depression despite taking several medications, Hester Eckert committed suicide. According to his colleagues, Eckert, who was always dour, became even more glum. The next year, Pres's father died in his arms. It all took a toll, and Eckert became more and more withdrawn.

The business wasn't going well either. Sales were not keeping pace with those of competitors, and in 1955, orders for IBM's 700 series of computers exceeded UNIVACs for the first time. IBM, with its legendary sales force and renowned hand-holding after the sale, was moving to dominate computing. Big Blue had scrambled to catch up in electronic computing, without Eckert and Mauchly, and it worked. The firm hired the usual suspects for help and even signed John von Neumann as a consultant in 1951. IBM was outspending Remington Rand on research and development, and soon its models were matching UNIVAC's capabilities. The battle was already won.

Faced with such competition, the Sperry Corporation bought Remington Rand in 1955, naming the Philadelphia computer unit the Remington Rand UNIVAC division of Sperry Rand Corp. Later, it became known as Sperry UNIVAC.

The computer industry, created by Presper Eckert and John Mauchly, had overtaken its founders. They had made many mistakes; Eckert bore much of the blame in the eyes of some colleagues. "Eckert was very smart, very aggressive, very entrepreneurial," said Chuan Chu, who started working for Eckert on the ENIAC project and stayed with him at EMCC, then had a long career in computing at Honeywell Corporation. "But he failed completely. He was unrealistic. He had wrong judgment."

How Sperry Stacked Up Against IBM

By the end of the 1960s, the computer industry was known as "IBM and the seven dwarfs." In 1965, IBM had 65 percent of the computer market. The seven dwarfs had 34 percent. Of the seven, Sperry Rand was the largest at 12 percent, followed by Control Data at 5 percent, Honeywell and Burroughs at 4 percent each, then General Electric, RCA, and NCR, all at about 3 percent.

But Sperry did have something IBM didn't have—the ENIAC patent. On February 4, 1964, after lengthy study and numerous challenges, including a rival application from Bell Labs for a machine only designed on paper and not actually constructed, Mauchly and Eckert received the patent on the computer, no. 3,120,606. That allowed Sperry Rand, which bought the patent rights along with the Eckert-Mauchly Computer division of Remington Rand, to collect royalties from every other computer maker. Had the patent been issued in 1947, when the application was filed, its normal seventeen-year life would have expired in 1964. Instead, the patent was issued in 1964, taking it up to 1981, a period of far greater computer sales. Sperry's market share may have been declining, but it was in a position to control the fast-growing computer industry into the 1980s.

Pioneers often suffer the hardships while paving an easier path for those who follow their trail. So it was with Eckert and Mauchly. The Eckert-Mauchly Computer Corporation had sparked the computer revolution, but the creators had little to show for it. The only thing that still bore their name was the original ENIAC patent. That was something they thought could never be taken away.

CHAPTER EIGHT

Whose Idea Was It, Anyway?

IBM didn't plunge into computing because it foresaw the future of the computer. Instead, this was a defensive action, designed, at the least, to protect IBM's dominance of office machinery *should* these new computers catch on. If they did catch on, they would be a way for IBM to diversify beyond its punch-card machines.

When he took over in 1924, Thomas Watson Sr. had set the company, which had grown out of Herman Hollerith's punch-card business, on a crusade to dominate every line of office machine. After the slight by Howard Aiken at the Mark I inauguration, and the failed attempt to hire Eckert and Mauchly in 1946, Watson recruited Wallace Eckert, a Columbia University electronics expert (and no relation to Presper), and established the Watson Computer Laboratory at Columbia. Electronic components were seeping into

IBM's most basic punch-card multipliers, and the company, whose computing efforts were driven by Watson's son, had taken on a major role building calculators for the government. But it was the launch of UNIVAC that had shifted the company into computing research overdrive.

"I thought, 'My God, here we are trying to build Defense Calculators, while UNIVAC is smart enough to start taking all the civilian business away!' I was terrified," Watson Jr. recalled in his autobiography.

IBM struggled with early scientific computers but then started to catch up with a new product for offices, the Model 650. The 650 was weak compared with UNIVAC but compatible with existing punch-card equipment in offices. That was a major selling point. UNIVAC used magnetic tape, a much better storage medium than punch cards, but the new technology required customers to convert all their card-punch records onto tape. IBM let them use all their existing cards, so the 650 was sold as an easy transition into computing. The 650 even fit into the same space as punch-card readers, and customers felt more comfortable with it. IBM quickly became a growing force in digital computing.

But the company had another reason to be eager for diversification into the new field of computers. In 1952, the U.S. Justice Department had filed an antitrust suit against IBM for its control of the office machines market. As a result, in 1956 IBM signed a consent decree that forced it to sell machines, not just lease them, and divest some punch-card operations. The settlement also forced IBM to sell competitors a license to build compatible machines. That was a key issue. If competitors couldn't sell products that would work with IBM office machines, they would likely be shut out completely from many offices.

Sperry Takes on IBM and the Other Six Dwarfs

Following the consent decree, Remington Rand wanted complete access to IBM patents. IBM figured that in return, from Remington, it should get access to ENIAC and UNIVAC computer patents that were still pending. The two couldn't strike a deal, and both sued each other. In 1956, one year after Sperry bought Remington Rand, Sperry and IBM settled their patent swap. They managed to keep the complete terms of their deal secret, for fear it would provoke more antitrust problems. What it amounted to was a cross-licensing deal under which IBM agreed to pay $10 million over an eight-year period, plus a $1.1 million down payment, for computer royalties—once Sperry received the ENIAC patent.

That was no small matter. Bell Labs had challenged the validity of the patent, as had IBM. IBM had gone so far as to track down Mauchly's old friend John V. Atanasoff to explore Mauchly's trip to Iowa and see if it was possible

Mauchly had gleaned the idea from Atanasoff. But IBM dropped its pursuit after signing the cross-licensing deal. In 1962, U.S. District Judge Archie Dawson upheld Eckert and Mauchly's claim, saying American Telephone and Telegraph, the parent of Bell Labs, had failed to produce evidence of "prior public use." That cleared the way for the issuance of the ENIAC patent in 1964.

Patent in hand, Sperry turned its guns on the other six dwarfs. In 1967, negotiations for royalties from Honeywell Corporation reached an impasse, and both firms sued each other, literally running to different courthouses trying to be the first to file. Sperry sued Honeywell for patent infringement in Washington, D.C., where patent litigation was frequently argued, while Honeywell sued Sperry in its home city of Minneapolis. Honeywell accused Sperry of demanding "discriminatory royalties" and of creating a "virtual monopoly" in the computer business with its cozy cross-licensing deal with IBM. Honeywell's suit in Minnesota was stamped fifteen minutes ahead of Sperry's patent-infringement claim in Washington. The District of Columbia circuit decided its docket was a lot fuller than the docket in Minneapolis, and therefore it was more than willing to let Minneapolis have the consolidated case. (Another Minnesota firm, William Norris's Control Data Corporation, also sued Sperry, but it was decided the Honeywell suit would go first.)

Even with the home-field advantage, Honeywell knew

it faced an uphill battle trying to contest the ENIAC patent, which had already been upheld by one federal court. Finding a judge to overturn another court was unlikely unless new ammunition could be found. Honeywell thought it probably had that in the once-secret cross-licensing deal. IBM paid only $11.1 million for rights to the ENIAC patent, but now Sperry was demanding $20 million from Honeywell, even though IBM's computer sales were sixteen times larger than Honeywell's. Sperry Rand had been seeking a total of $150 million from the six other dwarfs, even though all of them combined were less than one-third of IBM's size in the computer market. The only problem with the case was that the IBM deal was struck in 1956. It was 1967, and a judge would most likely rule that the statute of limitations had expired. Honeywell had to find something else new to cloud the ENIAC patent—something that would discredit Eckert and Mauchly.

New Ammunition for Honeywell's Case

Honeywell's lawyers got lucky. Henry L. Hanson, general counsel for Honeywell's patent division, happened to be a classmate of an Iowa State electrical engineering graduate named R. K. Richards, who had written an obscure book about computer development in which he mentioned the work of John Atanasoff at Iowa State—Mauchly's old friend.

Hanson mentioned the book to the Washington attorneys Honeywell had hired for the case. With the help of Iowa State, they tracked down Atanasoff, who happened to be in suburban Washington, just minutes from the lawyers' office.

Atanasoff was perfect for Honeywell's case. There was a paper trail of collaboration between Mauchly and Atanasoff, and, from a distance, his machine and ENIAC appeared to be something of the same ilk. What's more, Atanasoff never had done anything to protect his interests, so if the court was looking for a reason to strike down the Sperry patent, Atanasoff was a harmless avenue—he could not claim a patent himself. In other words, a judge might not want to take the patent and award it to someone else, undoing a long history and leaving the same sword hanging over the young industry. But a judge could use Atanasoff simply to invalidate the patent. The first court hadn't known of Atanasoff because IBM settled before ever revealing him. So the Minneapolis judge could overturn the earlier ruling based on new information, which would be preferable to saying the first judge was just plain wrong.

John V. (J. V.) Atanasoff had worked for several years on his computer at Iowa State with a graduate student named Clifford Berry. Searching for funding in the late 1930s, he took his idea to IBM, which turned a cold shoulder, and consulted with MIT and Bell Labs. Before Mauchly's visit to Iowa State in the summer of 1941, Atanasoff had already visited the U.S. Patent Office in Washington and hired his own Chicago patent attorney. He even was featured in a January 15, 1941, Des Moines *Tribune* story, with a picture, under the headline "Machine Remembers."

> AMES, IA—*An electrical computing machine said here to operate more like the human brain than any other such machine known to exist is being built by Dr. John V. Atanasoff, Iowa State College physics professor.*

But Atanasoff abandoned the project during World War II when he moved to Washington and went to work in the Naval Ordnance Laboratory. After the war, he stayed on with the research lab working on various defense projects, then went out on his own and formed his own consulting company. He ultimately sold the company to Aerojet General and was a retired millionaire when lawyers came calling in 1967.

The ENIAC patent was a sweeping document that made some 148 claims by the time it was fully amended and revamped. It claimed ENIAC was "the first general purpose automatic electronic digital computing machine known to us." That wording seemed to be a hedge against any competing claim from Atanasoff, whose machine was just a special-purpose machine capable of solving only one kind of problem. Yet what the patent really amounted to, by the time it was finally issued in 1964, was the rights to the system—to the computer itself. By stretching so far, the ENIAC patent was now vulnerable. Atanasoff had a "computer"; Eckert and Mauchly had a "computer." Which came first? The Atanasoff.

The machines were completely different, of course, like the difference between a bicycle and an automobile, though both are wheeled modes of transportation. In terms of basic design, the two machines were polar opposites. The Atanasoff machine was designed as a serial machine; information could progress only in straight-line order through one channel. ENIAC was a parallel machine; numbers and instructions could be routed around the machine through multiple channels, and multiple calculations could be done at the same time. ENIAC was programmable; the Atanasoff was not. That's a major, fundamental difference when talking about computer systems. Atanasoff's machine was data insensitive; it plodded ahead regardless of results, more like a calculator than a computer. Only ENIAC had the ability to do problems with a conditional branch—the If . . . then statement. Whereas ENIAC could automatically take the results of one computation and apply it to another, Atanasoff's machine could only spit out one result at a time. An interim finding had to be manually carried back to the start of the machine and fed in again.

"That was the fault of Atanasoff's machine," Mauchly said after Atanasoff resurfaced. "He had to stop and punch more buttons for everything he did. It was not only special-purpose, but could only run at human speed."

The structures of the two machines were entirely different, as well. ENIAC had a clock to govern its operations. The Atanasoff had no clock, so its internal operations couldn't be coordinated; once the problem was set, the machine went like a runaway train all the way to the end. The Atanasoff converted numbers to base 2; ENIAC ran in base 10. The two employed different logic in their circuits. ENIAC had the novel "counting circuits" Eckert had come up with that used tubes simply as on-off switches; the Atanasoff counted by measuring accumulated voltage inside a tube. ENIAC stored numbers in its accumulators, which could also perform math functions, while the Atanasoff had a rotating drum pocked with capacitors that stored numbers in the electrical charge of the capacitors.

That drum, in fact, was the key to the Atanasoff machine. What had initially attracted Mauchly, all parties agreed, was Atanasoff's claim at their first meeting that he could build his computer very cheaply. Mauchly, who wanted something to help him with his weather problems, knew he had to use vacuum tubes to get the speed necessary to solve the complicated equations. But he also knew building a big machine with vacuum tubes would be too expensive for him to do on his own. Atanasoff had told him he used tubes, and it was cheap. The reality was he used capacitors, which were cheap, and employed only a few tubes in the machine, mostly as amplifiers. The problem with that design was that capacitors are slow; they don't let you take advantage of the speed of the tubes. So for Mauchly's purposes, they weren't the answer. Atanasoff's bicycle didn't have the internal combustion engine; Mauchly wanted an eight-cylinder muscle machine with four wheels. Indeed, Mauchly said later he tried to convince Atanasoff to use flip-flops and scaling circuits, to no avail.

As a result, the two machines were worlds apart in speed. ENIAC was blistering fast, operating at 100,000 cycles per second. Atanasoff's machine was limited by the rotation of the memory drum, which went around sixty times a second.

What's more, Atanasoff's machine was never fully operational. Historians say the computing part did work, but there was a problem in the input and output mechanisms, which actually used electrical charges from capacitors to burn holes in cards, thus recording the numbers. The charred cards were then read by a card-reading machine. Atanasoff admitted the problems created errors once every 10,000 to 100,000 computations. In fact, because the machine was never fully operational, Atanasoff never was able to file his patent application.

"There was no resemblance to ENIAC at all," said Jack Davis, an engineer on the ENIAC team, in an interview. "I never heard of Atanasoff until I was called to testify."

ENIAC was developed three years after Mauchly's visit to Iowa, and Mauchly's technical know-how had rapidly progressed after he teamed up with Eckert. But had Atanasoff's gadget given Mauchly his idea? Was ENIAC derived from Atanasoff's machine? Atanasoff muddied Mauchly's claims, and that's exactly what Honeywell needed.

Mauchly and Atanasoff Square Off in the Court

The trial opened June 1, 1971, in Minneapolis before Judge Earl Larson, who was to render a verdict from the bench. Judge Larson heard testimony four days a week for several months from a total of seventy-seven witnesses. Eighty others were included by deposition, and lawyers produced a total of 32,654 exhibits, including a thick book describing the nineteenth-century work of Charles Babbage. The trial lasted 135 days, and testimony ended on March 13, 1972. The transcript ran 20,667 pages.

Even though Atanasoff had called his machine the Atanasoff Computing Machine, he and Honeywell's lawyers started referring to it as the Atanasoff-Berry Computer, or ABC. The new name was a tip of the hat to Cliff Berry, to whom Atanasoff had agreed in 1941 to give 10 percent of the royalties he received from the machine. From a lawyer's perspective, it made the Atanasoff creation sound simple, elementary—as if it was indeed the first computer.

What was in ENIAC that came from ABC? Atanasoff testified that he thought there were four unique concepts in his machine: regenerative memory, logic circuitry, serial calculation, and use of the base-2 system. ENIAC, however, had an entirely different form of memory and different logic circuitry. It was parallel, not serial, and it used base 10.

In short, there was nothing in ENIAC that Atanasoff could point to and identify as a copy of his work. He even testified that it was difficult to trace any specific parts in ENIAC back to the ABC. Still, he believed the overall concept of an automatic electronic digital computer was derived from ABC. The concept, after all, was what the ENIAC patent had so broadly tried to claim. It was the idea of building an electronic, digital computer that Atanasoff claimed Mauchly had stolen from him. Atanasoff claimed Mauchly had never thought about digital circuits until he visited him in Iowa, that Mauchly was still working solely on analog machines.

Mauchly denied that, saying he had been working on elementary digital devices long before he went to Iowa. While at Ursinus College, he had built a crude digital cryptologic machine that he unsuccessfully tried to sell to the military for coded messages before the war. He said he had built digital counting circuits, too, such as a little binary counter that looked like a railroad-crossing signal, and he had receipts from companies for vacuum tubes he had ordered before he visited Iowa. There was the September 27,

1939, letter to Supreme Instruments Corporation in Greenwood, Mississippi, inquiring about tubes, and another letter the same day to a different company in which he used the same "electrical calculating machine" terminology. There was also Mauchly's letter to a friend talking about an "electric computing machine" that would use vacuum tubes. All of this showed he was already onto the notion of a digital computer before paying his visit to Atanasoff.

Carl Chambers, an engineering colleague, remembered in an interview that Mauchly had made a "little computing device . . . [that] used neon tubes as trigger circuits. And he'd done some simple arithmetic work on the little desk set-up, using those triggers." Mauchly's primary electronic circuits using vacuum tubes had even been displayed years later at Pfahler Hall at Ursinus College, though Sperry's lawyers never presented them at the trial.

"This was not a flash from heaven, a full-blown device without any prior suggestion as to how anybody could do anything," Mauchly testified.

But John Mauchly didn't make a very good witness. He was defensive and forgetful. His health was worsening, and he was also very evasive. Later, he claimed Sperry's lawyers had encouraged him to remember as little as possible, but Mauchly probably even took that a step too far. He made a huge mistake by obfuscating the facts of his Iowa visit.

At first, Mauchly said he spent only ninety minutes

with Atanasoff's machine and did not get "any great detail or understanding to what Dr. Atanasoff had in mind." Actually, he spent several days with Atanasoff.

Mauchly arrived at Atanasoff's house in Ames, Iowa, on June 13, 1941—a Friday. He and Atanasoff went to the lab over the weekend to look at the machine, but Atanasoff was reluctant to fire it up until Clifford Berry was available Monday. Besides, only parts of the machine were functioning. The two spent considerable time together. Mauchly stayed until Wednesday, when he headed back to Philadelphia after receiving word that he had been accepted at the University of Pennsylvania's special summer course in electronics organized by the military.

Students of Atanasoff's testified that during Mauchly's visit they saw him with his sleeves rolled up nosing around the circuitry with Clifford Berry. Atanasoff had even let Mauchly read the machine's design documents.

Atanasoff testified that Mauchly was "ecstatic" at what he saw. Mauchly said he was deeply disappointed, calling the Atanasoff computer a pile of junk.

"This thing," he said, "is a mechanical gadget which uses some electronic tubes in operation, but it's still restricted in speed and not what I was interested in from the point of view of electronic speed gadgets. . . . it wasn't electronic in whole." He testified he went to Iowa with the same attitude he took to the World's Fair in 1939, to Stibitz's demonstration at Dartmouth. He was a sponge trying to soak up all new ideas, but he claimed he found no new useful ideas in Iowa. Once he saw the nature of Atanasoff's machine, he claimed, "I no longer became interested in the details."

A Smoking Gun?

The case largely boiled down to one scientist's words against the other. To bolster Atanasoff's view that Mauchly was indeed intensely interested in his machine, and that it was of use to him, lawyers used a series of letters back and forth between the two. The letters seemed to contradict Mauchly's after-the-fact view of the importance of the visit, and thus bolstered Atanasoff's credibility.

The two first met after Mauchly gave a speech in December 1940 on how he had solved some weather problems with an analog computing device he had built. Atanasoff had approached Mauchly to tell him of his own computing-machine work. Mauchly first wrote to Atanasoff on January 15, 1941, asking how his plans were working out, and saying that the idea of a visit to Iowa "grows on me." His main concern in the letter was the cost of the circuits; that Atanasoff's circuits cost two dollars per digit sounded "next to impossible," he said.

The answer to the two-dollar riddle, of course, was that Atanasoff wasn't using vacuum tubes as the primary electronic components; he was using his semimechanical rotating drum, pimpled with cheap, simple capacitors. So Mauchly's claim of disappointment had some validity, though it wasn't reflected in later letters.

Atanasoff responded to Mauchly's first letter by telling him that he was hunting for grant money and had visited MIT, where he got "a rather complete picture of calculating machine activities in the country." He said he hoped Mauchly would pay a visit because Atanasoff had lots of things to ask him about, including "computing machines of all kinds."

Letters continued back and forth, with Mauchly inquiring in May 1941 why Atanasoff's project hadn't become a defense project, since the military had a huge need for computing power and Atanasoff had consulted with people involved in national defense research. The answer, years later, was that no one had seen the potential for Atanasoff's machine, and for good reason. It would only handle one kind of mathematical problem, even if it did work.

After Mauchly's visit to Iowa in June 1941, the letters continued. Upon his return, Mauchly wrote to say he had had lots of ideas about computing machines during his drive back across the plains. "If any look promising, you may hear more later," he said.

On June 28, Mauchly wrote to H. Helm Clayton, a meteorologist friend, telling him of his trip to Iowa State. "My own computing devices use a different principle," Mauchly said, seemingly bolstering his testimony.

But then a letter was produced that did Mauchly in. It began "Dear J. V.," and it was dated September 30, 1941. In it, Mauchly delivered what was painted in the trial as a smoking gun.

"A number of different ideas have come to me recently about computing circuits—some of which are more or less hybrids, combining your methods with other things, and some of which are nothing like your machine. The question in my mind is this: Is there any objection, from your point of view, to my building some sort of computer which incorporates some of the features of your machine?" Mauchly wrote. Only recently hired at the Moore School, Mauchly even asked if the Atanasoff design "were to hold the field against all challengers . . . would the way be open for us to build an Atanasoff Calculator (a la *Bush* analyzer) here?"

The impact of the letter was devastating. Even though the ENIAC proposal came a year later, developed with new ideas of different ways to build a computer, the letter undercut everything Mauchly had tried to claim about the irrelevance of Atanasoff's work. Mauchly had buried himself.

He didn't march back to Penn and steal Atanasoff's idea; many of the letters to Atanasoff and others clearly showed that Mauchly was going in a far different direction and had been thinking of digital electronic circuits before he ever went to Iowa. But then again, the September 30 letter definitely indicated he hadn't thought ABC was simply a "pile of junk."

Colleagues testified that Mauchly never mentioned

Atanasoff or his trip to Iowa. This, too, was somewhow viewed as incriminating, even when the colleagues argued Mauchly never mentioned him because he never contributed anything. Only John Mauchly could explain why he never brought up the trip. But Mauchly had lost much of his credibility in court.

Why Wasn't Atanasoff Mentioned in the ENIAC Patent Filing?

U.S. patent law, 35USC Section 101, says: "Whoever invents or discovers any new and useful . . . machine . . . or any new and useful improvement thereof, may obtain a patent therefore. . . . A person shall be entitled to a patent unless . . . he did not himself invent the subject matter sought to be patented. . . . The applicant shall make oath that he believes himself to be the original and first inventor."

Were Eckert and Mauchly the "original and first" inventors of the computer? Honeywell argued they had a duty to disclose Atanasoff's work, and claimed that Eckert and Mauchly instead had conspired to hide that work.

That claim seemed to be a stretch, especially since Eckert mentioned Atanasoff's memory system in an article published in a scientific journal in 1953—before the Richards book or any other publication made mention of Atanasoff. "Probably the first example of what might generally be termed regenerative memory was developed earlier than 1942 by Atanasoff in Iowa," Eckert wrote. "He used a drum with many capacitors mounted on it. . . . Unfortunately, his development was interrupted by the war and never completed."

Eckert and Mauchly argued there was no reason to mention Atanasoff in the ENIAC patent filing because there was nothing—neither a memory drum nor anything else—that was derived from Atanasoff's device.

Patents are a tricky matter. If you look hard enough, you will find that almost every invention includes a glimmer of suggestion of someone else's work. That's the foundation of knowledge—we take what we know, and advance it. Somewhere we draw the line and say yes, that is something new, something novel, an "idea" worth crediting. There was no question that Mauchly had learned from Atanasoff, but Mauchly and Eckert also learned from RCA, which helped with the circuits, and from other calculating machines. Inventors aren't expected to be ignorant. ENIAC, everyone agreed, was a unique system, born of fresh ideas and past knowledge. Atanasoff was now claiming it was his theory Mauchly had stolen, but you can't patent theory; you can only patent something that works. Sperry assumed all of this worked in its favor. There was no way that Mauchly's visit to Iowa could undo all the work that had become ENIAC.

Honeywell's Groundbreaking Use of Computers in the Trial

During the trial, there was a special irony. Honeywell relied heavily on a computer to help organize its case. Testimony, depositions, and exhibits were all cataloged into a computer, and lawyers could quickly and easily cross-reference claims from witnesses. When Mauchly told his account of an incident, the computer helped lawyers instantly produce past statements contradicting a particular point. It was a groundbreaking use of computing power, breaking down a mountain of information into neat little piles for the attorneys and for the judge, and it caught Sperry's lawyers flat-footed. The use of a computer during the trial also showed the judge one practical application of the invention, and its coming importance in all areas of business—that is, if Sperry could be prevented from strangling competition with its outrageous royalty demands.

Four days after the trial ended, Honeywell blistered the court with a 5,000-page computerized brief summing up its arguments with reams of references and materials. Sperry's lawyers filed a short, traditional brief only in support of their claim that Honeywell had infringed on the ENIAC patent, and then put together a 300-page narrative reply to the Honeywell attack. That drew another 5,000-page computer dump from Honeywell, coupled with a more traditional narrative brief of about 400 pages.

"We found the Honeywell computerized brief format quite unpalatable," Sperry attorney William Cleaver said in a December 14, 1972, letter while both sides awaited Judge Larson's decision. "We worked diligently to digest its vast content and to reply to it in some reasonable fashion."

The Judge Hands Down His Decision

It was nearly seven months after testimony ended before Judge Larson handed down his decision. The ruling ran more than two hundred pages itself. The judge took a simple tack, concluding that the ENIAC patent application was filed six months too late, so it was invalid. A patent application has to be filed within one year of first "public use" of the invention. The ENIAC patent application was filed June 26, 1947, but Judge Larson held that the Los Alamos experiment run in December 1945 constituted a public use. It wasn't that the experiment was done publicly—it was top secret—it was that the machine had been turned over to its customer, the army, and essentially was being operated by people other than its inventors. It was no longer under the "surveillance" of Eckert and Mauchly; they no longer had control of the machine.

Sperry had claimed that the ENIAC was still "experimental" at that stage, that the Los Alamos exercise was simply a shakedown cruise. But the judge said the work for Los

Alamos was actually quite normal use. Besides, the public unveiling of ENIAC had also come more than one year before Eckert and Mauchly finally got their application filed. Douglas Hartee, the British scientist, even commented on ENIAC in an article in *Nature* magazine dated April 20, 1946, once again, more than a year before the application was filed. Hartee himself used the machine in 1946. And Larson also noted that Eckert and Mauchly were trying to sell ENIAC—actually UNIVAC, but much the same technology—before the application was filed. His finding: "The claimed invention disclosed in the ENIAC was on sale prior to the critical date."

Larson also zeroed in on the infamous *First Draft* report. Even though it covered EDVAC, it amounted to an "enabling disclosure of the ENIAC," the judge said. Von Neumann had somehow struck again. Larson called the document an "anticipatory publication," predating the ENIAC patent application by two years.

Judge Larson also attacked Sperry for trying to amend its patent application as late as 1963. He called it an example of "late claiming"—expanding a patent's claims to inflate its scope and extend its life. In effect, the judge said, Sperry had tried to tuck subsequent contributions to the field under the ENIAC patent and extend the patent's life with the amendments. "Deliberately extending the expiration of a monopoly is a serious violation of the Constitution and the patent laws," Larson wrote.

As far as Atanasoff was concerned, Judge Larson found him more credible than Mauchly; he also found the evidence persuasive that Mauchly had taken Atanasoff's idea and made it his own. Atanasoff's "breadboard model established the soundness of the basic principles of design," the judge said. Mauchly had been "broadly interested in electrical analog calculating devices" before visiting Atanasoff, Larson said, "but had not conceived an automatic electronic digital computer."

The judge even noted that Mauchly had invited Atanasoff to ENIAC's first public demonstration. Mauchly, of course, had been working as a consultant for Atanasoff at the Naval Ordnance Lab and knew of his interest in computing machines four years earlier, so it would seem logical that he extend an invitation. Instead, Judge Larson interpreted the invitation as a tip of the hat to a major influence.

"The subject matter of one or more claims of the ENIAC was derived from Atanasoff, and the invention claimed in the ENIAC was derived from Atanasoff," the judge said. "Eckert and Mauchly did not themselves first invent the automatic electronic digital computer, but instead derived that subject matter from one Dr. John Vincent Atanasoff."

It was a brutal ruling for Eckert and Mauchly. Having seen their ideas credited to von Neumann, and having lost their own company, the two men were now stripped of their last pride. Not even ENIAC was theirs anymore.

At the same time, Judge Larson's ruling was somewhat contradictory. He also found that Eckert and Mauchly alone were indeed the inventors of ENIAC, and that the ENIAC patent claims did not rest on Atanasoff's work. He called ENIAC a "pioneering achievement" and said that Mauchly may in good faith have believed he did not derive anything from Atanasoff. Indeed, the judge noted that in a September 1944 diarylike memo on ENIAC's development, Mauchly had mentioned Atanasoff, saying, "I thought his machine was very ingenious, but since it was in part mechanical (involving rotating commutators for switching) it was not by any means what I had in mind."

Did the Judge Do the Right Thing?

Whether Judge Larson was correct or mistaken has been a subject of passionate debate within the circle of computer historians. But the meat of the judge's 248-page order has been overlooked and misunderstood. After all, the suit was Sperry versus Honeywell, not Atanasoff versus Mauchly. They were just the bit players in a bigger drama of dominance in the computer field, then called Electronic Data Processing, or EDP.

At the time, IBM was regarded as the two-ton elephant that sat on anyone it wanted. The view was particularly prevalent in Minneapolis, where not only Honeywell but also Control Data, in adjacent St. Paul, complained bitterly about the state of the computer industry. IBM commanded what appeared to be exclusive use of the patent on computers—the ENIAC patent—through the once-secret cross-licensing deal it had made with Sperry in 1956. In essence, the No. 1 and No. 2 players in the industry appeared to be trying to freeze everyone else out. That was the bigger issue—not the personal glory of either Mauchly or Atanasoff.

As Honeywell's lawyers had hoped, Judge Larson was clearly angry with Sperry. In his ruling, the judge said the 1956 deal, details of which he found the two companies had tried to keep secret, was stifling to the industry, noting that in no other industry was technological development so important. The agreement was a "technological merger" of two companies that controlled 95 percent of the industry, but that "merger" had not been subjected to any antitrust regulatory approval. It was a de facto exclusive deal, he said, because the rest of the industry was cut out. Without access to those patents, Honeywell, for example, couldn't make peripheral equipment that would work with UNIVAC gear or IBM gear. Businesses had to go totally with IBM and UNIVAC or totally with Honeywell, whose designs had to be completely different. Most chose the safety of IBM and UNIVAC.

"I find that the cross-license and exchange of technical information agreement was an unreasonable restraint of trade and was an attempt by IBM and Sperry Rand to strengthen or solidify their monopoly in the EDP industry," the judge wrote. Sperry Rand, he said, "conspired and agreed with IBM to prevent Honeywell from obtaining access to any IBM or SR patent licenses and know-how."

As a result, what Judge Larson really wanted to do was open up the computer industry, let Honeywell, Control Data, and others have access to patent licenses and know-how. It was the proper thing to do legally, given that Sperry and IBM had engaged in a conspiracy to monopolize the industry. It also was clearly the right thing to do for the nation. The adolescent computer industry was beginning to have an enormous impact on companies and government, and its development was vital to the U.S. economy. Judge Larson recognized the need for competition. If he didn't act, Sperry and IBM would have had a stranglehold at least until 1981—nearly eight more years.

But he couldn't just come down hard on Sperry and IBM, which wasn't even a party directly involved in the suit. The 1956 agreement between Sperry and IBM was subject to a four-year statute of limitations, so the judge said he was barred from acting on that.

How best, then, to open up the industry? The answer was simple: invalidate the ENIAC patent—and take away Sperry's club. If there was no patent, everyone would have free access to computer technology. It was the perfect solution to the problem. It solved Honeywell's problem, it punished Sperry for its transgressions, and it cleared the way for competition, an enormously important development. Atanasoff, von Neumann, the Los Alamos test, and all the other events simply gave Judge Larson a way to do the right thing and make sure his ruling would hold up on appeal. The judge didn't even bother awarding Honeywell damages or even attorney's fees. He simply broke the cartel.

It had happened before. American inventor George Selden had been awarded the patent on the automobile in the United States in 1895. Several early manufacturers licensed by Selden formed an association and took over the patent, threatening to control the industry. Henry Ford, however, refused to acknowledge the patent and sued. A federal court ruled that the patent, while valid, covered only cars with two-cycle engines. Most cars, including Ford's, by then used four-cycle engines. Competition could emerge because the court found a way to make it happen.

Many commentators argued that Judge Larson didn't have all the facts, so he came to the wrong conclusion on the ENIAC patent, or he didn't understand the technology, so he didn't really know what he was doing. But he knew exactly what he was doing: He was breaking a monopoly, freeing computer development to take off in the 1970s and 1980s. Mauchly and Eckert were just innocent bystanders caught in the cross fire.

"I think the judge had to decide that way because if that patent had gone to UNIVAC it would have meant they had a patented monopoly on all the future of the computer industry," Grace Murray Hopper, the computer software pioneer, said in a 1976 interview. "And I think he took any straw that he could find to see that they didn't get it. I think it was a political and financial and future-of-the-nation decision, so to speak."

Reactions to the Legal Decision

Once the judge made his decision, Atanasoff became an instant celebrity. The forgotten inventor of the computer—what a great story that was! The rather harshly worded decision in Atanasoff's favor played well in newspaper accounts. ENIAC had faded from public consciousness in the 1970s, and even though the computer was becoming a more important tool in society, its invention was a blurred nonevent even before the Atanasoff controversy. Most people probably thought IBM invented the computer. Now a quirky retired physicist was declared the inventor three decades after the fact by a federal judge. *Incredible!*

The decision certainly caught the computer science field by surprise. As might be expected, everyone took sides.

Herman Goldstine, carrying on the legacy for the now-deceased von Neumann, wrote a well-regarded book detailing the historical development of the computer. The first half of the book leads up to ENIAC, and the second half, by his own admission, is autobiographical. Goldstine said that Atanasoff never amounted to much, that his chief contribution "was to influence the thinking of another physicist who was much interested in the computational process, John W. Mauchly."

Arthur Burks, the onetime No. 3 man on ENIAC, also stepped into the fray. Burks had gone to Princeton for a time with von Neumann and ultimately settled back at the University of Michigan, where he had started his career as a philosopher. He and his wife wrote an article in the *Annals the History of Computing* siding with Atanasoff, and the article ultimately led to a book: *The First Electronic Computer: The Atanasoff Story.*

The Burkses said they believed Mauchly stole the basic principles of ABC but went so far beyond that with ENIAC that he and Eckert "felt safe in appropriating those basic concepts because Atanasoff was showing no interest." The Burkses added that Eckert and Mauchly "were greedy, for fame and fortune, and did not want to acknowledge any prior inventor."

In fact, the second half of that last sentence applied to Arthur Burks as well. He had thought his own contributions to the logical design of ENIAC merited inclusion on the patent application.

Nevertheless, the Burkses said they thought the judge had been correct "in this finding of prior invention by Atanasoff," saying they had come to agree that critical ideas had been passed to Mauchly, and that "Atanasoff's computer was the first electronic computer." They qualified that by saying Atanasoff was indeed special purpose, and ENIAC was general purpose.

Despite the support from several factions, the Atanasoff story faded quickly. The verdict had come down at the height of the Watergate scandal, with which the nation was preoccupied. Besides, the members of the computer industry, with the exception of Goldstine and Arthur Burks, had stuck with Eckert and Mauchly because ENIAC was significant and well known, and ABC was totally insignificant and obscure. Academics continued to refer to them as the inventors of the computer. Grist Brainerd, of all people, even rose to Eckert and Mauchly's defense. He was part of a faction suggesting that Judge Larson did not have enough "background" to understand the machines and realize how completely different they were. Eckert and Mauchly continued to refer to themselves as the inventors of the computer, as did many historical accounts of the computer field.

An Iowa Journalist Stirs Up the Controversy Again

Despite the fact that the court's decision had made Atanasoff the inventor of the computer, he and his wife were convinced he wasn't getting his due from the computer field or the public at large. He decided to fight back.

His wife called a friend from school who was a well-known journalist, Clark Mollenhoff, Washington correspondent for the *Des Moines Register.* She told him the story and asked for advice. Mollenhoff realized she was giving him a great story. In Iowa, no one had picked up on the ruling giving an Iowan credit for inventing the computer. Mollenhoff dug into the tale and published both a newspaper story that triggered enormous interest in Iowa and a subsequent book, *Atanasoff: Forgotten Father of the Computer,* published in 1988 by Iowa State University Press.

Mollenhoff went even farther than Judge Larson and Arthur Burks had gone, saying Mauchly "stole Atanasoff's electronic digital computer ideas and falsely claimed to be the true inventor of those concepts." He painted Atanasoff as a naive inventor who was duped by a devious colleague he thought to be honorable. Mauchly had sweet-talked his way to Iowa, an "eager and enthusiastic supporter and admirer," Mollenhoff wrote, "who was anxious to help in any way to give him full credit for any of the ideas derived from the Atanasoff-Berry machine." Then Mauchly turned around and pilfered from Atanasoff. Atanasoff, Mollenhoff said, was too hurt and disappointed to challenge the ENIAC patent, even during the lengthy patent review.

The story grew truly bizarre when Atanasoff, in Mollenhoff's book, began raising accusatory questions about Clifford Berry's death in 1962, which had been ruled a suicide. Mollenhoff said Atanasoff "wondered again if his untimely and mysterious death was in any way related to Berry's renewed interest in 1962 in digging into the links between the Atanasoff Berry Computer and ENIAC."

That insinuation seemed especially outrageous since Atanasoff, in 1972 interviews at the Smithsonian, sixteen years before Mollenhoff's book, had recounted the depths of Berry's depression shortly before his death. Berry had been in a car accident that left him in substantial pain and then was hurt again in a second car accident. Atanasoff had visited his former student in California and found him in bad shape. He was drinking, depressed, his marriage was strained, and he was unhappy with his work, Atanasoff said. Berry was under the care of a "psychological advisor" for stress, and was "consuming a considerable amount of alcohol," Atanasoff said. Berry took a job on Long Island, New York, traveling across the country without his wife, and moved into a garage apartment. It was there that Berry was found in his bed with a plastic bag over his head. Numerous liquor bottles were found in the garage after his suicide, and his blood plasma level at the time of death showed him to be intoxicated.

What Was the Real Atanasoff Story?

A bigger mystery than Clifford Berry's death was why Atanasoff—if he truly considered himself the rightful inventor of the computer—remained silent about his invention for so many decades until he was contacted by Honeywell's attorneys. Not only did he not challenge Eckert and Mauchly, he never tried to remind anyone of his original device. Why? Even in 1940, long before ENIAC, Atanasoff, like Eckert and Mauchly, had understood the commercial implications of computers. He had, after all, contacted IBM about his invention. He had his own patent attorney. On April 6, 1940, he even wrote to Remington Rand, saying he could build "a computing machine" that could replace tabulating machines with a faster, cheaper electronic device. And he had visited Harvard and MIT, and knew his way around computing circles.

"I was always surprised that Atanasoff did not surface as a continuing contributor," Isaac Auerbach said. "I never understood this because this is rather atypical of a guy who really thinks he's got a great idea. . . . He was a drop-out. I never understood why he left the business."

Atanasoff said he abandoned computing because of the war, then went on to other things. He said he had simply assumed Mauchly had used his own ideas. Mauchly, with whom Atanasoff had frequent contact at the Naval Ordnance Lab, told him ENIAC was classified and he couldn't talk about it, Atanasoff said, and he never knew what was actually in ENIAC until lawyers brought him the patent many years later. "I wasn't possessed with the idea I had invented the first computing machine. If I had known the things I had in my machine, I would have kept going on it," he said in an interview with the *Washington Post.*

Could that be?

Actually, Atanasoff, like Mauchly, proved to have a talent for stretching his case. Recently uncovered records show Atanasoff was far more involved in computing than he let on in later years, and he in fact was very knowledgeable about ENIAC.

Since the army, its fiercest rival, had a computer, the navy decided it had to have a computer as soon as it heard about ENIAC. The navy turned to someone with a long background in computer development to lead it into the computer age: Atanasoff.

Records show that Atanasoff represented the Naval Ordnance Laboratory at a navy conference on computing research on February 27, 1946. The conference took place less than two weeks after the ENIAC public unveiling, and ENIAC was a prime topic of discussion. Though he claimed he left the computing field when he left Iowa, Atanasoff was actually named to lead a computer-building program at the Naval Ordnance Laboratory and given a $300,000 budget for the task, a far cry from the $5,000 he scraped together to help fund ABC.

And even though he later claimed he never had access to ENIAC, records at the Library of Congress and Smithsonian Institution indicate just the opposite. Atanasoff had intimate access to ENIAC, and he even hired von Neumann as a consultant to help with his computer-building program for the navy. Atanasoff's deputy, Robert D. Elbourn, attended the famous "Moore School Lectures," which were cosponsored by the Naval Ordnance Laboratory, and which were basically a how-to course in construction of an ENIAC-type computer. Another Naval Ordnance Laboratory worker under Atanasoff, Calvin N. Mooers, even gave one of the forty-eight lectures in that Moore School series. His topic: the computer development program under Atanasoff.

Before all that, Goldstine had even sent a copy of the well-traveled *First Draft* report to Atanasoff at the Naval Ordnance Laboratory, records at the Library of Congress indicate. John Vincent Atanasoff, it turns out, was very much in the vortex of computer development and had a wealth of knowledge about ENIAC before him, even though he later claimed the contrary. A letter even turned up showing that in 1941, Mauchly suggested Atanasoff apply for a position at the Moore School to return to computer development work. (Mauchly's invitation could hardly be seen as the work of a man trying to hide a theft of ideas from the original source.) Despite that familiarity, Atanasoff didn't claim any of his ideas had been appropriated by Mauchly.

Atanasoff's computer program failed for the navy. People who worked with him said Atanasoff looked at ENIAC and didn't like it, so he went off in a more ambitious, if misguided, direction. He tried to build a computer using a television picture tube as an electrostatic storage device. The project never got very far, and the navy pulled the plug, reclaimed the unspent money, and shifted its computing efforts elsewhere. Atanasoff later claimed the naval computer project was underfunded, even though he spent only $15,000 of the $300,000.

Robert Elbourn, Atanasoff's deputy who had attended the "Moore School Lectures," said in a 1971 interview with the Smithsonian that the navy computer crew never even got as far as designing an architecture for their machine. "For some reason, he was very reluctant to tell us anything about that machine he had designed," Elbourn said of Atanasoff. "Evidently, he felt that the state of the art had changed and we ought to make a fresh start. He didn't want to influence us, and cause us to do things the way he had done them." Mooers, also on the navy computer project, said it "was doomed by poor management."

Atanasoff's recollection of that period was quite different. In his 1972 interviews at the Smithsonian, Atanasoff claimed he could have resigned from the Naval Ordnance Laboratory after the war to pursue computing projects but didn't want to. Computers were a "great interest to me," he said. "The rest of the story is rather peculiar. I did not, at that time, in 1946, realize the great importance of the work which I had done in the over-all computing machines. I didn't realize that, as a matter of fact, the concepts which I had were the best that there were in the computing art. I didn't realize I was the best man—strictly speaking, I perhaps had the best grasp of elements of computing in 1946, as good as anyone. Historically speaking, the ideas which I conceived were more advantageous than those of others, but I did not myself realize that that was true. In retrospect I realize it now."

The contention, of course, ignores his hiring of a patent attorney in 1940, his visits to IBM and Remington Rand, and his failure to build a computer when given a large budget and access to the best consultants. His deputy, Elbourn, even disputed Atanasoff's claim that Mauchly had hidden details of ENIAC from him. In his role as a consultant, Mauchly, Elbourn recalled, dropped in every week, "and quite a little time was spent listening to him describe the progress on the ENIAC. He also discussed with him [Atanasoff] . . . their plans for the ENIAC."

Mauchly said he sometimes asked Atanasoff what became of his machine when he was consulting for him. "He didn't seem to want to talk about it," Mauchly said.

"That's One Comment on History"

Judge Larson's decision was never appealed. Sperry, having spent more than $1 million, likely realized it had dodged a bullet on the antitrust issue and was content to simply call it a draw. Fighting for the ENIAC patent was dangerous, risky, and expensive. Eckert and Mauchly asked for their rights to the patent back so that they might appeal, but Sperry refused.

Mauchly was openly heartbroken, but Eckert, as usual, kept his bitterness largely to himself. A friend of Eckert's said recently that Eckert probably understood what the judge had accomplished with his decision, and since he often found himself at odds with Sperry executives and had in fact opposed the 1956 deal with IBM, he kept quiet. "Maybe he understood the implications of the decision on the industry. Maybe that's why he never challenged it," the friend said.

Eckert did take a more historical view of his fate, comforting himself with the knowledge that many inventors didn't receive proper credit until events were digested with the wisdom of time. He argued that he and Mauchly were no different from Edison or the Wright brothers, whose inventions were technically not the first in their class but the best. "Many people built light bulbs before Edison, and some had been manufactured five years earlier. What Edison invented was a system with an improved light bulb and various features and stuff," Eckert told historian Nancy Stern in 1980. "Atanasoff had no system. We built a system that worked. His was not an invention by definition of the patent office. Ours was."

In a 1991 speech in Tokyo, Eckert said: "The work by

Dr. Atanasoff in Iowa was, in my opinion, a joke. He never really got anything to work. He had no programming system. He tried for a patent and was told the work he had done was too incomplete to get a patent. A competitor in a patent suit convinced what in my mind was a very confused judge to believe Atanasoff's story, even though it had no real relation to the case at hand."

"Mauchly and I achieved a complete workable computing system. Others had not," Eckert said. "If Edison is the inventor of the incandescent lamp it would appear that by the same yardstick Mauchly and I are clearly the inventors of the computer."

Mauchly could never talk about the subject as dispassionately. In a videotaped interview shortly before his death, Mauchly recalled the gala twentieth anniversary celebration of the Association of Computing Machinery, which was held in Washington. Atanasoff, living nearby in Frederick, Maryland, never attended, and in fact that day had met with Honeywell's lawyers for the first time. "He didn't come to the big computer meeting. He never had come to a computer meeting, but he turned up in a lawsuit later," Mauchly said, choking back tears with a forced smile. "So that's one comment on history, you might say."

In 1989, as the Smithsonian Institution readied an exhibit on computer development at the National Museum of American History, researchers prepared to credit Eckert and Mauchly with inventing the computer. Word reached Iowa congressman Neal Smith, who complained the honor belonged to Atanasoff. The politically sensitive Smithsonian backed off.

Today at the Smithsonian's permanent exhibit, Atanasoff's picture is mixed into a display of computer pioneers called "Origins of the Electronic Computer." The Smithsonian says Atanasoff "built the first electronic computer" but notes it was a special-purpose machine that was never fully operational. Nearby, a huge display of ENIAC, including some of the original units, captures enormous interest from visitors. On a television screen, Pres Eckert proudly and meticulously recounts the creation of the giant marvel.

"This is part of ENIAC," a tourist recently told his teenage, computer-savvy son on their first visit to the Smithsonian. "This is part of the first computer."

"Talk about making history!" the teen gasped. "Who built it?"

EPILOGUE

So Much Has Been Taken Away

We still can't precisely predict the weather, but someday we will. Sometime in the future, John Mauchly's dream will become a reality. And when it happens, he most likely won't be remembered. Just as most people have already forgotten that Mauchly and Eckert invented the first true computer and founded the first computer company.

They never got rich from their invention: All told, Eckert and Mauchly each received about $200,000 to $250,000 over many years. Eckert had been independently wealthy, but for Mauchly, money was always a struggle.

Mauchly left Sperry Rand in 1960 and formed a company called Mauchly Associates that developed a computerized construction manager. Long before the desktop computer and before Compaq Computer Corporation and others started building "luggable" computers, Mauchly actually built a portable computer in a suitcase to demonstrate his construction management software.

The company became known as Scientific Resources, and it went into hospital design, including the use of computers in health care and refinery management in Algeria. But the personal computer was still years away, and communication between computers was still difficult. Once again, John Mauchly was too far ahead of himself with a good idea, and there was no market. The company did well enough to go public, but it ultimately failed and Mauchly's stock ended up worthless.

"I'm not a good salesman," he admitted. "I haven't seen how to sell some of my ideas. The selling of the original ENIAC to the Army Ordnance was a purely fortuitous thing based on the fact that there was a war going on, there was a need there and somehow I had gotten into the Moore School—the right place at the right time. It's a big game of chance. That time I happened to win, and the world happened to win."

His third company, organized in 1966, was called Dynatrend Incorporated. With five employees and a leased IBM Model 1130, Dynatrend did advanced research into computing applications. The firm handled programming and subcontracted the actual number crunching. As always, Mauchly was thinking big. He was never tied down by inhibitions or conventions and he spent his time studying questions like encryption for electronic funds transfer and

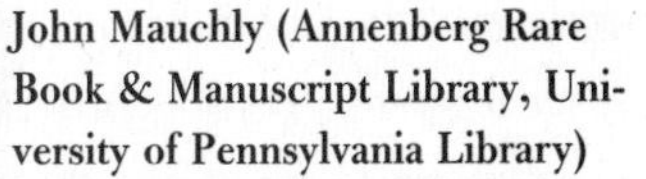

John Mauchly (Annenberg Rare Book & Manuscript Library, University of Pennsylvania Library)

electronic mail. Unfortunately for him, those kind of developments were still ten to twenty years away. Like his other ventures, Dynatrend never got very far.

"The biggest surprise to me was the development went so slowly," Mauchly said of computing. "It seemed to me that it was an obvious direction here, it had been laid out and a lot of people ought to take hold of this thing right away and the development should have gone faster than it actually did."

Dreaming proved not to be enough, even for the inventor of the computer. John Mauchly went broke. To keep his family's house and old farm in Ambler, Pennsylvania, he made a deal with a developer to subdivide some of the land

and build houses on it. Even that didn't work smoothly; the developer got tied up in a zoning battle, slowing the project. So it went. Lucky to keep his house, John Mauchly, hat in hand, went back to Sperry after Judge Larson's ruling and rejoined the company as a consultant in 1973.

He never did much for the company, but he was never far from computers. In 1975, Mauchly was one of the first to order a new computer from Radio Shack called the TRS-80. He had two, actually, one upstairs and one downstairs. He often called Fort Worth, Texas, to talk to Radio Shack engineers. Most of the time, he tried to use the TRS-80 to run weather-forecasting programs.

"He was a dreamer. Always thinking up new things," says Kay Mauchly Antonelli, his second wife.

Despite the business failings, the Mauchlys lived a fine family life in suburban Philadelphia. Each Sunday, the family—which eventually included two sons and five daughters—had a ritual of doing the *New York Times* crossword puzzle together. The wallpaper in the dining room was a map of the world, and Mauchly the teacher was quick to deliver a geography lecture. Books were everywhere—the house had a thirty-four-foot-long library.

The controversy never seemed to ease. Grist Brainerd resurfaced in 1975 to claim in a speech to the Society for the History of Technology in Washington, D.C., that *he* wrote out the original ENIAC proposal by hand "after consultation with the others who might be on the project." He

Mauchly and Eckert (The Charles Babbage Institute)

said he chose Mauchly, because of his earlier memo, and Eckert, because of his circuitry expertise, to work on *his* project. It was just one more indignity for Mauchly.

Later that year, Mauchly, whose health always seemed on the verge of disaster, suffered a brain aneurysm and was hospitalized for a lengthy period. He recovered, but not fully. He became more bitter, mostly about von Neumann. But the mention of Atanasoff could bring tears to his face.

"Things that were so simple, and the result of natural business exploration in those days, can be so twisted and colored by lawyers and those who have different motives," he wrote in his diary on September 20, 1977. "So much has been taken away . . ."

Mauchly died January 8, 1980, at the age of seventy-two. Pres Eckert cried all night when he heard the news, and he delivered the eulogy at Mauchly's funeral. He even had to stop for milk at an ACME store in Ambler on the way to the service to calm his sick stomach.

"They made a great team," said Eckert's second wife, Judy. "They were very different. But they were soul mates."

Pres had remained a vice president at Sperry. He met Judy in 1957 singing in a church choir, and they lived in exclusive Gladwyne, Pennsylvania, in a large house on a hill, where neighbors included the Annenbergs.

By the early 1960s, Eckert was advocating giving up on mainframe computers and plunging instead into cheap machines that could sit on a desktop. Once again laying out the future of computing, he talked about using computers for keeping business records, making decisions, controlling machines, analyzing data, distribution of goods, automatic accounting, and bookkeeping.

Even then, Eckert advocated an open programming language, instead of proprietary systems, so that customers can have free choice in suppliers, "which would keep costs lower for consumers." He predicted that there would be computers capable of holding the information of a complete encyclopedia, and that students could be connected to teaching computers, allowing each student to learn at his or her own pace and "free our teachers to handle the more creative aspects of education." Computers would find broad use in medicine, Eckert said, but manufacturers had to deliver high-performance machines at low cost.

"This is not quixotic pipe-dreaming by an over-imaginative computer man," he said in a 1962 interview. "But take my word for it. Compared to what is going to come, computers today are in the dark age of their lifetime."

Sperry didn't listen, and Eckert grew more and more out of favor. The company, which has since renamed itself UNISYS and is still based in suburban Philadelphia, stripped outspoken Eckert of his responsibilities. His only functions for the company became glad-handing customers and delivering speeches. After the verdict, there wasn't a lot of demand for that, either.

Eckert retired and picked up in his basement where he had left off, designing new things and plunging headlong into circuits and electronics. Often he muttered, "The world's not fair." Judy Eckert recalled, "It was always, 'This damn luck. . . . If it weren't for this damn luck . . .' "

One obsession was audio speakers. Eckert bought hundreds and took them apart, figuring out their construction and testing to see if they lived up to advertised specifications. He had always been fascinated by sound waves—the EDVAC memory was sound waves in mercury tanks. When the performance of speakers didn't match the manufacturer's claims, Eckert, stubborn as always, called the company to complain. "He always figured most people did not know what they're doing," friend Thomas Miller said.

To test his speakers, Eckert often dug holes in his well-kept yard, lowering the speaker and measuring frequencies without any interference. He always graphed the results and eventually blew out the speakers to see how much it took to exceed their limits. Then he carefully replaced the rectangle of grass sod.

"One day a neighbor came by and asked him what he was doing. 'Oh, I'm testing my tweeter,' he said," Judy Eckert recalled. "She said to me, 'How can you stand it?' I said, 'You get used to it.' "

Another obsession was his seventy-two-foot yacht, the *Miss Laura.* The boat came with radar, naturally, but it was not good enough, so Eckert redesigned the radar to his own specifications. He tried to eliminate all moving parts with electronic devices throughout the boat and had a dockside warehouse full of parts.

Every time a new electronic gadget came on the market, Pres Eckert had to have one so he could take it apart and see what was going on. Tom Miller loaned him a videocassette recorder once, which Eckert returned after a few days. It never worked quite right after the loan, Miller said. Shortly before his death, Eckert took apart a compact disc player, and also drove to Atlantic City to buy a slot machine so he could find out how that worked.

Pres Eckert in 1964, receiving an honorary degree from the University of Pennsylvania (The University of Pennsylvania Archives)

"He had a belief that anything could be solved. It was just a question of cost," said Miller.

In all, Eckert had eighty-seven patents to his name, including invalidated ENIAC. He had also worked secretly for the State Department for a number of years, traveling on a diplomatic passport on various projects. He always kept a bag packed, his wife said, and one day in the early 1960s he was summoned to examine some computer equipment the United States had "obtained" from the Soviet Union. Eckert was gone for several days, and all he would say upon his return was that the Russians knew nothing about computers. On January 20, 1969, the last day of the Johnson administration, Pres Eckert—alone—received the Medal of Honor from LBJ.

Over the years, Eckert had refused to appear on panels with Goldstine or Brainerd, or even be in the same room with Goldstine. But later in life, the two old pioneers, still revered by young turks who were carrying on their legacy, got together for lunch after they ran into each other at the taping of a Public Broadcasting Corporation special called "Giant Brains." To the surprise of their families, Eckert and Goldstine buried some of their bitterness and kept in touch.

Pres Eckert died of leukemia on June 3, 1995, at the age of seventy-six. His death came within eleven days of the death of John V. Atanasoff, who had lived to be ninety-one. The *Washington Post* ran an obituary of Atanasoff but not of Eckert.

In 1996, the United States celebrated the fiftieth anniversary of the unveiling of ENIAC. Vice President Al Gore gave the keynote speech at the University of Pennsylvania, and IBM sponsored the first match between Garry Kasparov and the Deep Blue computer, which Kasparov won. Man still was smarter than computers, but the match made a great story. The event was carefully worded to not offend anyone in Iowa, but the bottom line was unmistakable: This was a celebration of the invention of the computer. It was a fitting, and honest, tribute to the pioneers of the computer age.

The next year, the army, hoping to remind everyone that it had paid for the first computer, hosted a fiftieth-anniversary celebration of ENIAC's arrival at Aberdeen. As army misfortune would have it, the event coincided with a sex scandal in the ranks at the Aberdeen Proving Ground, and the commanding officer was absent because he had been summoned to the Pentagon.

On a bitterly cold Maryland day, with the wind whipping in off the bay, troops paraded at Aberdeen to honor Herman Goldstine, still fit, sharp, and dapper at age eighty-three. Television cameras were there, but they focused on women marching in camouflage fatigues, not on Goldstine. As the honoree rode standing up in the back of a Humvee surveying the troops, a young soldier read his army "biography."

Goldstine, the announcer said, was "put in charge of the Moore School."

The stretch drew sighs and laughs from Judy Eckert, Kay Mauchly Antonelli, and several ENIAC veterans who were in the review stand huddled under blankets.

"And he has been proclaimed as the founder of the information age," the army official belted out.

"Oh my God," exclaimed Kay. "When will it end?"

When will it end? What happened to Eckert and Mauchly doesn't seem to compute. They showed the world that electricity could be used to do calculations. They were the first to produce a working machine that not only fits the definition of what a modern computer is but also is easily traceable as the father of later machines. After ENIAC, they were largely responsible for the first computer to incorporate the stored-program concept. They invented the first commercial electronic digital computer in the United States.

Even IBM's chess-playing Deep Blue is a descendant of Eckert's and Mauchly's creations.

More important, Eckert and Mauchly indisputably were the founding fathers of the commercial computer industry. The Eckert-Mauchly Computer Corporation was the first computer company in the world. Its founders sparked a new technology that changed the world.

Despite all that, Eckert and Mauchly were blinking lights themselves. Their careers were on-again, off-again affairs, flip-flops. Their role in history, too, has faded in and out as the creation of the computer has been rewritten and the industry has developed. In time, like the blinking lights attached to ENIAC, they will undoubtedly become a more permanent part of the structure of computing.

There is a fundamental postulate in science that when an idea's time has come, it will blossom in many places simultaneously, fertilized by the work of predecessors. "If I have seen further than other men," Isaac Newton said, "it is because I have stood on the shoulders of giants."

So it was with computing machines. Construction of new devices was under way in many laboratories around the world. Information was shared, orally and in written form. Visits were made; theories and ideas were traded. At the opening lecture of the Moore School course in the summer of 1946, George Stibitz, the czar of Bell Labs, paid homage to the work of Babbage, Bush, Aiken, and himself. "Each invention rests upon prior inventions and upon techniques which were already developed," Stibitz noted.

That was certainly true of ENIAC. Yet at the same time, ENIAC was unique. It broke away from conventional computing techniques. ENIAC took ideas from electromechanical computing machines and made them electronic. Accumulators were patterned after adding machines. Function tables were a machine version of paper mathematical tables. Charles Babbage had the dream more than a century earlier, and he laid out the blueprint for the world. But Pres Eckert and John Mauchly were the ones who made the computer a reality.

"People who change the world stand facing two ways in history," Maurice Wilkes, the British computer pioneer, said in a tribute to Eckert. "Half their work is seen as the culmination of efforts of the past; the other half sets a new direction for the future. This is especially true of Eckert. The ENIAC was the culmination of earlier efforts."

But today, their triumph is tangled in words and qualifications. A plaque on the Moore School at the corner of Thirty-third and Walnut in Philadelphia reads: "Birthplace of ENIAC, World's First Electronic Large-Scale, General-Purpose Digital Computer." But what is a "computer," in today's language, if not electronic, if not digital, if not large-scale (powerful), and if not general-purpose?

Many played a role in the development of the computer, but Eckert and Mauchly were the ones who ultimately put it all together. Yet the plaque on the Moore School building doesn't even mention them. It could well say, as Marshall Ledger suggested in the *Pennsylvania Ga-*

zette alumni newspaper in 1982, "Here were hopes and expectations quashed and egos gored. Here was trust shattered, credit appropriated or slighted, and history muddled into bitterness."

Despite all that, Mauchly still believed the record would one day be corrected and his contribution would be appreciated. Historians, he hoped, would be kinder to him than his contemporaries were. "History changes on how it treats various people," Mauchly said in an interview shortly before his death. "History is certainly going to change its point of view about me and Eckert and lots of other people. We think it'll change with respect to who did what so as to reflect the part we really played in the invention of the computer."

Eckert was even more unflagging, able to draw self-satisfaction from the explosion of computers in everyday life. "In spite of all the problems of the world, which I think the computer can help more than anything else, working on computers has always been a lot of fun and I feel very fortunate that I have been able to be a part of it," he said in a 1991 speech in Tokyo.

He ended that speech with the same line he used to end a speech at the Computer Museum in Boston, Massachusetts: "How would you like to have most of your life's work end up on a square centimeter of silicon?"

Notes

Introduction. The Thinking Man's Game

"It showed a sign of Intelligence": Garry Kasparov quoted by Janine Zuniga, Associated Press, May 7, 1997.

"Who invented the computer?": Bob Levey, *Washington Post,* May 29, 1998.

Chapter One. The Ancestors

"So what? Won't *we* know?": Herbert Kelleher, interview by author, Dallas, Texas, 1998.

There are several good sources on the early history of computing devices, especially Herman Goldstine, *The Computer from Pascal to von Neumann* (Princeton, N.J.: Princeton University Press, 1972) and *A History of Computing in the Twentieth Century,* edited by Nicholas Metropolis, Jack Howlett, and Gian-Carlo Rota (New York: Academic Press, 1980). Other valuable resources are Stan Augarten, *Bit by Bit,* Martin Campbell-Kelly and William Aspray, *Computer: A History of the Information Age* (New York: HarperCollins, 1996), Paul E. Ceruzzi, *Reckoners: The Prehistory of the Digital Computer from Relays to the Stored Program Concept, 1935–1945* (Westport, Conn.: Greenwood Press, 1983), and Howard Rheingold, *Tools for Thought* (New York: Simon and Schuster, 1985).

"Everywhere . . . there will be IBM machines": Watson quoted in Campbell-Kelly and Aspray, *Computer,* p. 48.

Chapter Two. A Kid and a Dreamer

In addition to interviews with family members, I relied on several accounts of John W. Mauchly's childhood, including Marshall Ledger in the *Pennsylvania Gazette,* October 1982. Mauchly's personal papers are housed at the University of Pennsylvania, Philadelphia, and include diaries, résumés, correspondence, and family budgets.

"Physicists—those were the boys": Mauchly, interview by Esther Carr, videotape recording, Ambler, Pa., 1978.

"I have a sort of stubborn streak": ibid. interview.

"Here at last was the course that I wanted": ibid.

Material on Presper Eckert's childhood comes from interviews with colleagues, friends and family members, personal papers and mementos in the family attic, and Eckstein's biographical paper in *IEEE Annals of the History of Computing.*

"In class, he was always testing the teachers": Jack Davis, interview by author, Quakerstown, Pa., May 1, 1997.

"If you're going to come to class": Harold Pender quoted by S. Reid Warren, in interview by Nancy Stern, tape recording, Philadelphia, Pa., October 5, 1977, OH 38, Charles Babbage Institute, University of Minnesota, Minneapolis.

The Osculometer is recounted by Herman Lukoff in *From Dits to Bits: A Personal History of the Electronic Computer* (Portland, Oreg.: Robotics Press, 1975).

Eckert's first patent, "Light Modulating Methods and Apparatus," dated May 19, 1942, contained in personal papers, property of Judy Eckert.

"Pres always had on a white linen shirt": Kathleen Mauchly Antonelli, interview by author, Aberdeen, Md., November 14, 1996.

"We would sit around on lab tables": Mauchly, Carr interview.

Mauchly's seven-page proposal, "The Use of High-Speed Vacuum Tube Devices for Calculation," and Brainerd's attached note, are found in the University of Pennsylvania archives, Philadelphia.

"None of us had much confidence in Mauchly": Carl Chambers, interview by Nancy Stern, tape recording, Philadelphia, Pa., November 22, 1977, OH 7, Charles Babbage Institute, University of Minnesota, Minneapolis.

Chapter Three. Crunched by Numbers

One important account of ENIAC development and early computer theory comes from the "Moore School Lectures." Some of the lectures were tape-recorded, and some were reconstructed from shorthand notes taken by one student.

The account of the trip to Aberdeen was based on the author's interview with Herman Goldstine in Aberdeen, Md., on November 13, 1996, Goldstine's book, *The Computer from Pascal to von Neumann,* and Mauchly's retelling to Carr.

"John never set out to build a computer": J. Presper Eckert, interview by Nancy Stern, tape recording, Philadelphia, Pa., January 23, 1980, OH 11, Charles Babbage Institute, University of Minnesota, Minneapolis.

"Dr. Brainerd was pooh-poohing the idea": Mauchly diary entry labeled "Situation as of 9/10/44."

"At that time during the war": Lila Butler, interview by author, Aberdeen, Md., April 29, 1997.

"Simon, give Goldstine the money": Herman H. Goldstine, interview by author, Aberdeen, Md., September 13, 1996.

Chapter Four. Getting Started

"The smartest thing we did": Eckert quoted in *Pennsylvania Triangle,* March 1962.

"He spoke well": Jean Bartik, unpublished memoirs.

"but he wasn't much up on what was going on": Davis, author interview.

"It was hoped originally that counter circuits": Moore School of Electrical Engineering, University of Pennsylvania, "Report of a Diff. Analyzer," document sent to Ballistic Research Laboratory, Aberdeen Proving Ground, Md., April 2, 1943.

"If it was going to work": Eckert, Stern interview.

"People thought I was a nut": Eckert in undated videotaped interview, courtesy of Aberdeen Proving Grounds.

"A serious attitude prevailed": Lukoff, *From Dits to Bits.*

"I was scared to death of him": Bartik, unpublished memoirs.

"There wasn't a single one of the staff": Chambers, Stern interview.

"Eckert set the most exigent standards": Goldstine, *The Computer from Pascal to von Neumann,* p. 154.

"Eckert was always pressing everything": Davis, author interview.

"He'd get an idea": Jean Bartik, interview by author, Aberdeen, Md.

"He loved to talk": ibid.

"We were really stuck": Eckert, Stern interview.

"John had brilliant ideas": in J. Presper Eckert et al., interview by Nancy Stern, tape recording, Philadelphia, Pa., January 23, 1980, OH 11, Charles Babbage Institute, University of Minnesota, Minneapolis.

"Guess this is a job": Kathleen Mauchly Antonelli, interview by author, Ambler, Pa., April 29, 1997.

"Mauchly was a physicist": Herman H. Goldstine, interview by author, Aberdeen, Md., November 13, 1996.

"I didn't get along as well with Mauchly": ibid.

"Eckert, in recent conversations": Mauchly diary entry labeled "Situation as of 9/10/44."

"We were young": Goldstine, author interview, September 13, 1996.

Chapter Five. Five Times One Thousand

"We have finally done it": Ledger, *Pennsylvania Gazette,* October 1982.

"I was astounded": Antonelli, author interview, November 14, 1996.

"We saw answers glowing": Eckert quoted in *Pennsylvania Triangle,* March 1962.

"Neighboring units did not catch fire": Goldstine, author interview, November 13, 1996.

"The ENIAC was a son-of-a-bitch to program": Bartik, author interview, Aberdeen, Md., November 14, 1996.

"It was the most exciting work": Bartik, author interview, Ambler, Pa., April 29, 1997.

"Grist Brainerd is . . . quite introspective": Warren, Stern interview.

"In view of the fact": Eckert and Mauchly letter to Warren.

"We would talk": Nicholas C. Metropolis, interview by William Aspray, tape recording, Los Alamos, N.M., May 29, 1987, OH 135, Charles Babbage Institute, University of Minnesota, Minneapolis.

The ENIAC unveiling account is based on the War Department's press release, issued February 15, 1946, the University of Pennsylvania's invitation and menu, plus other documents in the University of Pennsylvania archives, including a transcript of a radio program aired on WCAV, Philadelphia, with Eckert, Mauchly, and Goldstine. Curiously, Mauchly said he wrote the script for the radio show.

"All kinds of things entered our minds": Mauchly, Carr interview.

"The War Department tonight unveiled": Associated Press, February 14, 1946, reprinted in the *Penn Paper,* February 6, 1986, p. 3.

The tale of the inquiry from the Russian government came from the *Pennsylvania Triangle,* March 1962.

"We did a lot of calculations": Joseph Chernow, interview by author, Philadelphia, Pa., November 12, 1996.

Chapter Six. Whose Machine Was It, Anyway?

The discussion of Goldstine's meeting with von Neumann is based on the author's interview with Goldstine, November 13, 1996.

"I was not familiar with great mathematicians": Eckert, Stern interview.

"He grasped what we were doing": Eckert, in Eckert et al., Stern interview.

"We constructed a machine": Eckert, Moore School Lectures (1946), vol. 9 in reprint series by Charles Babbage Institute, University of Minnesota, Minneapolis, 1985.

"You are right": Burks tells this story in a tape-recorded interview with Christopher Evans in Ann Arbor, Mich., 1976, OH 78, Charles Babbage Institute, University of Minnesota, Minneapolis.

"There was this kind of dichotomy": John W. Mauchly, interview by Henry S. Tropp, tape recording, February 6, 1973, no. 196, Computer Oral History Collection, Smithsonian Institution, Washington, D.C.

"The word 'hierarchy' was suggested": John W. Mauchly, interview by Christopher Evans for Pioneers of Computing (Science Museum, London), tape recording, Philadelphia, Pa., ca. 1976, OH 26, Charles Babbage Institute, University of Minnesota, Minneapolis.

"I remember saying 'no' ": Bartik, author interview, April 29, 1997.

"Von Neumann and Teller had one characteristic in common": Eckert, in Eckert et al., Stern interview.

The account of *First Draft* is based largely on the author's interview with Goldstine on November 13, 1996, as well as recorded interviews of Mauchly and Eckert.

"It appeared to be purely a characteristic of Goldstine": Mauchly, Tropp interview.

"[Goldstine] asked if this material could be mimeographed": Warren memo dated April 2, 1947.

"It was damning with faint praise": Mauchly, Carr interview.

"The invention of the acoustic delay line memory device": John W. Mauchly and J. Presper Eckert Jr., "A Progress Report on EDVAC," internal document, University of Pennsylvania, September 30, 1945.

"A great influence in all of this": Eckert, in Eckert et al., Stern interview.

"They didn't have the patience": Warren, Stern interview.

"Not everything in there is his": Goldstine, *The Computer from Pascal to von Neumann,* p. 191.

"While the placing of the EDVAC report": ibid.

"[Von Neumann] appreciated at once the possibilities": Wilkes, "A Tribute to Presper Eckert." *Communications of the ACM* 38, no. 9 (September 1995).

"He spoke with a forked tongue": Eckert, Stern interview.

"In my opinion, we were clearly suckered": J. Presper Eckert Jr., keynote speech, Imperial Hotel, Tokyo, Japan, April 15, 1991.

"He took every credit": Mauchly, Carr interview.

"From my point of view": Warren, Stern interview.

"Far as I know, Eckert came up with the . . . concept": Davis, author interview.

"Von Neumann had very little impact": Brad Sheppard, telephone interview by author, April 30, 1997.

"It is clear that the stored-program concept": Metropolis and Worlton, "A Trilogy of Errors in the History of Computing." *Annals of the History of Computing.* Vol. 2, No. 1, January, 1980.

"[Use of the term 'Von Neumann architecture'] has done an injustice": Campbell-Kelly and Aspray, *Computer,* p. 95.

"It is suggested": Colonel Paul N. Gillon to Moore School of Electrical Engineering, April 10, 1946.

"Herman was a pretty vindictive sort of personality": Eckert, in Eckert et al., Stern interview.

"The following were mentioned": Mauchly diary, Sunday, January 20, 1946.

"All people who wish to continue as employees": Irven Travis, interview by Nancy Stern, tape recording, Paoli, Pa., October 21, 1977, OH 36, Charles Babbage Institute, University of Minnesota, Minneapolis.

"The sense was Eckert was so dominant": Davis, author interview.

"[Eckert and Mauchly have to] certify you will devote": letter to Eckert and Mauchly from Dean Harold Pender, March 22, 1946.

"[The patent policy was] very, very naive": Warren, Stern interview.

"Perhaps it would have made a difference": Ralph Showers, telephone interview by author, May 1, 1997.

"They asked too late": Goldstine, author interview, November 13, 1996.

"When I die": Eckert quoted by Judy Eckert, interview by author, Aberdeen, Md., November 13, 1996.

Chapter Seven. Out on Their Own

Sources on post-ENIAC period: author's interview of Goldstine on November 13, 1996, Mauchly personal papers, Eckert interviews, Antonelli interviews, Lukoff's personal account in *From Dits to Bits,* as well as Electronic Control Company and Eckert-Mauchly Computer Corporation papers at the Hagley Museum and Library, Wilmington, Del.

"I said, 'Well, I happen to think' ": Eckert, in Eckert et al., Stern interview.

"We felt that the time was at hand": Mauchly personal letter to Fred Weiland, April 15, 1946, a copy of which was contained in his personal papers at the University of Pennsylvania archives, Philadelphia.

"We got together and we did this thing": Eckert, "Moore School Lectures."

"[she wished] they'd hurry up": Mary Mauchly letter to her mother, postmarked May 30, 1946, contained in Mauchly papers, University of Pennsylvania, Philadelphia.

Mary Mauchly's drowning was recounted in the *Philadelphia Inquirer*, September 9, 1946.

A stenographer took minutes of the patent meeting on EDVAC. A transcript is included in the ENIAC trial record.

ENIAC patent, no. 3,120,606, dated February 4, 1964, "Electronic Numerical Integrator and Computer," John Presper Eckert Jr. and John W. Mauchly, Philadelphia, Penn., assignors.

"Everything all day long was ideas": Isaac Auerbach, interview by Henry S. Tropp, tape recording, Washington, D.C., February 17, 1972, no. 196, box 3, file 5, Computer Oral History Collection, Smithsonian Institution, Washington, D.C.

"It was such an amazing sight": Earl Edgar Masterson, interview by William Aspray and Robbin Clamons, tape recording, St. Louis Park, Minn., June 30, 1986, OH 115, Charles Babbage Institute, University of Minnesota, Minneapolis.

"Eckert was always a believer": Auerbach, Tropp interview.

"The first thing he'd say": Davis, author interview.

"Life is very lonely": letter from Kay McNulty to Mauchly, September 24, 1947, contained in Mauchly's personal papers.

"Eckert and Mauchly were singularly unprejudiced": Grace Hopper, tape-recorded interview for Charles Babbage Institute, ca. 1976, OH 81, University of Minnesota, Minneapolis.

"The more I think about the situation": Mauchly memo to EMCC employees, February 5, 1948, contained in Sperry-UNIVAC Company Records, Hagley Museum and Library, Wilmington, Del., accession 1825, box 1, series 1, subseries I.

"They lacked a salesman": Brad Sheppard, author interview.

"Would you come in . . . ?": Isaac Auerbach, interview by Nancy Stern, tape recording, April 10, 1978, OH 2, Charles Babbage Institute, University of Minnesota, Minneapolis.

"When I dug into the contracts": Margaret Fox, interview by James Ross, tape recording, Minneapolis, Minn., April 13, 1983, OH 49, Charles Babbage Institute, University of Minnesota, Minneapolis.

"Literally, chalk and blackboard erasers flew about the room": Auerbach, Stern interview.

"The intensity was just horrendous": Davis, author interview.

"A great deal of personal animosity": Auerbach, Stern interview.

"Mauchly's wife was mysteriously drowned": Mauchly's FBI dossier is contained in Augarten's *Bit by Bit.*

"The truth is that it was never given the chance": Mauchly in March 22, 1978, letter to Dennis Cooper, Olympia, Wash., written on Radio Shack TRS-80 computer.

"[Mauchly was a] lanky character": Thomas Watson Jr. and Peter Petre, *Father and Son & Co.* (London: Bantam Press, 1990), p. 198.

"That was a pretty damn big achievement": Mauchly, Carr interview.

"[Designing and building a computer was] not of sufficient general interest": Eckert in text of undated speech titled "Review of the History of Computing," p. 4. Courtesy of Judy Eckert.

"Eckert was very smart": Chan Chu, telephone interview by author, May 10, 1997.

Chapter Eight. Whose Idea Was It, Anyway?

One by-product of the detailed litigation is that much of the early material was not only preserved but also assembled in a complete source. The documents create a valuable record of ENIAC issues as well as early computer development. There are actually three repositories of documents and transcripts from the litigation: the Charles Babbage Institute in Minneapolis, the University of Pennsylvania archives in Philadelphia, and the Hagley Museum and Library in Wilmington, Del. In addition, the Hagley collection has early company records from Sperry UNIVAC, including records from the Eckert-Mauchly Computer Corporation.

"I thought, 'My God, here we are trying to build Defense Calculators' ": Watson Jr., *Father and Son & Co.*, p. 227.

John V. Atanasoff details his pitches to IBM and Remington Rand, his trip to the U.S. Patent Office, and his work with a Chicago patent attorney in his interview with Henry S. Tropp, tape recording, Washington, D.C., February 18, 1972, no. 196, box 2, files 7, 9, 12, Computer Oral History Collection, Smithsonian Institution, Washington, D.C.

"An electrical computing machine": *Des Moines Tribune,* January 15, 1941. The story ran deep inside the paper next to dispatches headlined "2 Lodge Protests Over Billboard" and "Paralysis Aid Totals $1,490." A copy can be found in the University of Pennsylvania archives, Philadelphia.

"The ENIAC is extremely flexible": ENIAC patent, paragraph 8.

"That was the fault of Atanasoff's machine": Mauchly, Carr interview.

"[Mauchly had made a] little computing device": Chambers, Stern interview.

"This was not a flash from heaven": Mauchly testimony, under questioning by Henry Halladay, attorney for Honeywell.

"[Mauchly did not get] any great detail": Mauchly testimony.

"This thing is a mechanical gadget": Mauchly testimony, pp. 11, 829–30 in transcript.

"I no longer became interested in the details": Mauchly testimony.

Letters between Mauchly and Atanasoff are available from several sources, including the University of Pennsylvania archives, the collection of Mauchly's papers, as well as the trial record at Hagley.

"Probably the first example . . . of regenerative memory": J. Presper Eckert Jr., "A Survey of Digital Computer Memory Systems," *Proceedings of the Institute of Radio Engineers* 41 (October 1953): 1934–1406.

"We found the Honeywell computerized brief format": William E. Cleaver letter to Bartik, December 14, 1972.

"The claimed invention disclosed in the ENIAC": Larson, "Findings of Fact, Conclusions of Law and Order for Judgment," in ENIAC Trial Records, *Honeywell Inc. v. Sperry Rand Corp, et al.*, no. 4-67, civ. 138, finding 2.1.

"Deliberately extending the expiration of a monopoly": ibid.

"The subject matter of one or more claims of the ENIAC": Larson, ibid., finding 3.1.

"I find that the cross-license and exchange of technical information agreement": ibid., finding 15.25.

"I think the judge had to decide that way": Hopper interview.

"[Atanasoff's chief contribution] was to influence the thinking of . . . Mauchly": Goldstine, *The Computer from Pascal to von Neumann.*

"[Mauchly and Eckert] felt safe in appropriating those basic concepts": Alice R. Burks and Arthur W. Burks, *The First Electronic Computer: The Atanasoff Story* (Ann Arbor, Mich.: University of Michigan Press, 1988).

"[Mauchly] stole Atanasoff's electronic digital computer ideas": Clark R. Mollenhoff, *Atanasoff: Forgotten Father of the Computer* (Ames, Iowa: Iowa University Press, 1988), p. 4.

"[Atanasoff] wondered again if his untimely . . . death": ibid, p. 233.

"I was always surprised that Atanasoff did not surface": Auerbach, Tropp interview.

"I wasn't possessed with the idea": Atanasoff quoted in the *Washington Post,* January 13, 1974, in a story by W. David Gardner.

Atanasoff's presence at the navy conference on computing, in room 3116 at the Navy Department, is noted in Mauchly's diary for that day, February 27, 1946, and filed in the University of Pennsylvania archives, Mauchly papers collection, Van Pelt Library, series 2: box B: folder 10a. In a 1972 interview at the Smithsonian, he claimed the Naval Ordnance Laboratory project was underfunded.

Records on Atanasoff's $300,000 budget and failed naval computer project come from Calvin Mooers's account, contained in documents on the "Moore School Lectures." "Moore School Lecture" records at the Library of Congress also list Elbourn as an attendee, Mooers as a lecturer.

"For some reason": Robert D. Elbourn, interview by Richard R. Mertz, Tape recording, March 23, 1971, no. 196, box 6, file 10, Computer Oral History Collection, Smithsonian Institution, Washington, D.C.

"[The navy computer project] was doomed by poor management": Mooers to editors of Moore School Lectures (1946), reprinted series by Charles Babbage Institute, University of Minnesota, Minneapolis. Mooers said Mauchly had arranged for Mooers to lecture on Atanasoff's Naval Ordinance Laboratory project.

"[Computers were a] great interest to me": John V. Atanasoff, interview by Bonnie Kaplan, Tape recording, Washington, D.C., July 17, 1972, no. 196, box 2, file 14, Computer Oral History Collection, Smithsonian Institution, Washington, D.C.

"And quite a little time was spent listening to him": Elbourn, Mertz interview.

"He didn't seem to want to talk about it": Mauchly, Carr interview.

"Many people built light bulbs": Eckert, in Eckert et al., Stern interview.

"The work by Dr. Atanasoff was . . . a joke": Eckert, Tokyo speech.

"He didn't come to the big computing meeting": Mauchly, Carr interview.

Epilogue. So Much Has Been Taken Away

"I'm not a good salesman": Mauchly quoted in *Sperry UNIVAC News* 4, no. 3 (February 1980) from one of his last formal interviews.

"The biggest surprise to me was the [slow] development": Mauchly, Evans interview.

"He was a dreamer": Antonelli, author interview, April 29, 1997.

"[Brainerd wrote out the ENIAC proposal] after consultation with the others": J. Grist Brainerd, text of address to the Society for the History of Technology, Washington, D.C. October 17, 1975, University of Pennsylvania archives, Philadelphia.

"Things that were so simple": Mauchly diary, September 20, 1977.

"They made a great team": Eckert, author interview, November 13, 1996.

"[Teaching computers could] free our teachers": Eckert interview in *Pennsylvania Triangle,* March 1962.

"This is not quixotic pipe-dreaming": ibid.

"One day a neighbor came by": Eckert, author interview, April 28, 1997.

"He had a belief": Thomas A. Miller, interview by author, Gladwyne, Pa., April 28, 1997.

"People who change the world": Wilkes, "A Tribute to Presper Eckert."

"Here were hopes": Ledger, *Pennsylvania Gazette.*

"History changes": Mauchly, *Sperry UNIVAC News.*

"In spite of all the problems of the world": Eckert, Tokyo speech.

Bibliography

Books

Augarten, Stan. *Bit by Bit.* New York: Ticknor and Fields, 1984.

Bartimus, Tad, and Scott McCartney. *Trinity's Children.* New York: Harcourt Brace Jovanovich, 1991.

Burks, Alice R., and Arthur W. Burks. *The First Electronic Computer: The Atanasoff Story.* Ann Arbor, Mich.: University of Michigan Press, 1988.

Campbell-Kelly, Martin, and William Aspray. *Computer: A History of the Information Age.* New York: HarperCollins, 1996.

Campbell-Kelly, Martin, and Michael R. Williams, eds. *The Moore School Lectures.* Vol. 9 in the Charles Babbage Institute reprint series on the history of computing. Cambridge, Mass.: MIT Press, 1985.

Carpenter, B. E., and R. W. Doran, eds. *A. M. Turing's ACE Report of 1946 and Other Papers.* Cambridge, Mass.: MIT Press, 1986.

Ceruzzi, Paul E. *Reckoners: The Prehistory of the Digital Computer from Relays to the Stored Program Concept, 1935–1945.* Westport, Conn.: Greenwood Press, 1983.

Goldstine, Herman H. *The Computer from Pascal to von Neumann.* Princeton, N.J.: Princeton University Press, 1972.

Lukoff, Herman. *From Dits to Bits: A Personal History of the Electronic Computer.* Portland, Oreg.: Robotics Press, 1979.

Macrae, Norman. *John von Neumann: The Scientific Genius Who Pioneered the Modern Computer, Game Theory, Nuclear Deterrence, and Much More.* New York: Pantheon, 1992.

Metropolis, Nicholas, Jack Howlett, and Gian-Carlo Rota, eds. *A History of Computing in the Twentieth Century.* New York: Academic Press, 1980.

Mollenhoff, Clark R. *Atanasoff: Forgotten Father of the Computer.* Ames, Iowa: Iowa University Press, 1988.

Randell, Brian, ed. *The Origins of Digital Computers: Selected Papers.* Berlin: Springer-Verlag, 1973.

Rheingold, Howard. *Tools for Thought.* New York: Simon and Schuster, 1985.

Rhodes, Richard. *The Making of the Atomic Bomb.* New York: Simon and Schuster, 1986.

Ritchie, David. *The Computer Pioneers.* New York: Simon and Schuster, 1986.

Shurkin, Joel. *Engines of the Mind: A History of the Computer.* New York: Norton, 1984.

Stern, Nancy. *From ENIAC to UNIVAC: An Appraisal of the Eckert-Mauchly Computers.* Bedford, Mass.: Digital Press, Digital Equipment Corp., 1981.

Watson, Thomas Jr., and Peter Petre. *Father and Son & Co.* London: Bantam Press, 1990.

Wulforst, Harry. *Breakthrough to the Computer Age.* New York: Charles Scribner's Sons, 1982.

Articles, Papers, and Documents

Atanasoff, John V. "Computing Machine for the Solution of Large Systems of Linear Algebraic Equations." Unpublished paper. No. 196, box 8, file 26, Computer History Archives, Smithsonian Institution, Washington, D.C.

Auerbach, Isaac, chair. "Panel on Computer Development." Association of Computing Machinery, Twentieth Anniversary meeting, August 30, 1967. Transcript in no. 196, box 2, file 3, Computer Oral History Collection, Smithsonian Institution, Washington, D.C.

Brainerd, J. Grist. Text of address to the Society for the History of Technology, Washington, D.C., October 17, 1975. University of Pennsylvania archives, Philadelphia.

Burks, Arthur W. "From ENIAC to the Stored-Program Computer: Two

Revolutions in Computing." In *A History of Computing in the Twentieth Century,* ed. Nicholas Metropolis, Jack Howlett, and Gian-Carlo Rota. New York: Academic Press, 1980, 311–44.

Burks, Arthur W., and A. R. Burks. "The ENIAC." *IEEE Annals of the History of Computing* 3, no. 4 (October 1981): 310–99.

Burks, Arthur W., Herman H. Goldstine, and John von Neumann. "Preliminary Discussion of the Logical Design of an Electronic Computing Instrument." Report for Research and Development Service, Ordnance Department, U.S. Army. 2nd ed., September 2, 1947.

Chapline, Joseph. "The Second Miracle of Philadelphia." Unpublished essay. Newbury, N.H., May 4, 1995.

Clippinger, R. F. "A Logical Coding System Applied to the ENIAC." Report no. 673. Aberdeen Proving Ground, Md., September 29, 1948.

Costello, John. "As the Twig Is Bent: The Early Life of John Mauchly." *IEEE Annals of the History of Computing* 18, no. 1 (spring 1996): 45–50.

———. "The Little Known Creators of the Computer." *Nation's Business,* December 1975, pp. 56–62.

Dewees, Anne. "It's a Better World, Thanks to John." *Sperry UNIVAC News* (Sperry Corp., Blue Bell, Penn.) 4, no. 3 (February 1980): 1–8.

Eckert, J. Presper, Jr. "Eulogy for John Mauchly." Reprinted in *Sperry UNIVAC News* 4, no. 3 (February 1980): 6.

———. Personal papers. Courtesy of Judy Eckert.

———. "A Survey of Digital Computer Memory Systems." *Proceedings of the Institute of Radio Engineers* 41 (October 1953): 1393–406.

———. "Yesterday, Today, and Tomorrow." Keynote speech, Imperial Hotel, Tokyo, Japan, April 15, 1991.

Eckert, John Presper, Jr., and John W. Mauchly. "Electronic Numerical Integrator and Computer." U.S. patent no. 3,120,606, filed June 26, 1947, issued February 4, 1964.

Eckstein, Peter. "J. Presper Eckert." *IEEE Annals of the History of Computing.* 18, no. 1 (spring 1996): 25–44.

Electronic Control Company and Eckert-Mauchly Computer Corporation. Corporate papers. Hagley Museum and Library, Wilmington, Del.

ENIAC Trial Records. Depositions, complaints, transcripts, exhibits, briefs, and decision. *Honeywell Inc. v Sperry Rand Corp. et al.,* no. 4–67, civ. 138, Minn. Filed May 26, 1967, decided October 19, 1973.

Fritz, W. Barkley. "ENIAC—a Problem Solver." *IEEE Annals of the History of Computing* 16, no. 1 (spring 1994): 25–45.

——. "The Women of ENIAC." *IEEE Annals of the History of Computing* 18, no. 3 (fall 1996): 13–28.

Goldstine, Herman H. "Computers at the University of Pennsylvania's Moore School, 1943–1946." Jayne Lecture delivered January 24, 1991, and reprinted in *Proceedings of the American Philosophical Society* 136, no. 1 (1992).

Goldstine, Herman H., and A. Goldstine. "The Electronic Numerical Integrator and Computer (ENIAC)." *IEEE Annals of the History of Computing* 18, no. 1 (spring 1996): 10–16.

Grier, David. "The ENIAC, the Verb 'to Program,' and the Emergence of Digital Computers." *IEEE Annals of the History of Computing* 18, no. 1 (spring 1996): 51–55.

Hartee, D. R. "The ENIAC, an Electronic Computing Machine." *Nature,* October 12, 1946, 500–506.

Hopper, Grace M., and John W. Mauchly. "Influence of Programming Techniques on the Design of Computers." *Proceedings of the Institute of Radio Engineers* 41 (October 1953): 1250–54.

Infield, Tom. "Faster Than a Speeding Bullet." *Philadelphia Inquirer,* February 4, 1996.

Kempf, Karl. "Electronic Computers within the Ordnance Corps." Ballistic Research Laboratory, Aberdeen Proving Ground, Md., November 1961.

Kennedy, T. R. "Electronic Computer Flashes Answers, May Speed Engineering." *New York Times,* February 15, 1946.

Ledger, Marshall. "ENIAC." *Pennsylvania Gazette,* October 1982.

Marcus, Mitchell, and Atsushi Akera. "Exploring the Architecture of an Early Machine: The Historical Significance of the ENIAC Machine Architecture." *IEEE Annals of the History of Computing* 18, no. 1 (spring 1996): 17–24.

Mauchly, John W. "Mathematical Machines with Myths Concerning Their Makers, or Babbage vs. Gutenberg." Lecture at the National Bureau of Statistics Colloquium, Ambler, Pa., February 23, 1973.

——. Personal papers. Courtesy of Rare Book Collection, University of Pennsylvania Library, Philadelphia.

——. "The Use of High Speed Vacuum Tube Devices for Calculating." Unpublished memo. University of Pennsylvania, Philadelphia, ca. 1943.

Mauchly, John W., and J. Presper Eckert Jr. "A Progress Report on EDVAC." Internal document. University of Pennsylvania, Philadelphia, September 30, 1945.

Moore School of Electrical Engineering, University of Pennsylvania. "The ENIAC, Volume I. A Report Covering Work Until December 31, 1943." Document sent to the Ballistic Research Laboratory, Aberdeen Proving Ground, Md., ca. 1944.
———. "ENIAC Progress Report." Document sent to the Ballistic Research Laboratory, Aberdeen Proving Ground, Md., July 31, 1944.
———. "Minutes of Patent Meeting." Transcript produced by stenographer, ca. 1946.
———. "Report of a Diff. Analyzer." Document sent to the Ballistic Research Laboratory, Aberdeen Proving Ground, Md., April 2, 1943.
Petzinger, Thomas, Jr. "Female Pioneers Fostered Practicality in Computer Industry." *Wall Street Journal,* November 22, 1996.
———. "History of Software Begins with the Work of Some Brainy Women." *Wall Street Journal,* November 15, 1996.
Smith, Sharon. "Clash of the Titans." *Computer Weekly,* March 7, 1996.
Von Neumann, John. "First Draft of a Report on the EDVAC." Unpublished document. Moore School of Electrical Engineering, University of Pennsylvania, Philadelphia, June 30, 1945.
Weik, Martin H. "The ENIAC Story." *Ordnance,* January–February 1961.
Wilkes, Maurice V. "A Tribute to Presper Eckert." *Communications of the ACM* 38, no. 9 (September 1995): 20–22.
Winegrade, Dilys. "Celebrating the Birth of Modern Computing." *IEEE Annals of the History of Computing* 18, no. 1 (spring 1996): 5–9.
Zuniga, Janine. "IBM Scientist: Kasparov Could Have Played for Draw in Game 2." Associated Press, May 7, 1997.

Archived Interviews

Alt, Franz. Interview by Henry Tropp. Tape recording, September 12, 1972. No. 196, box 1, file 10, Computer Oral History Collection, Smithsonian Institution, Washington, D.C.
Atanasoff, John V. Interview by Bonnie Kaplan. Tape recording, Washington, D.C., July 17, 1972. No. 196, box 2, file 14, Computer Oral History Collection, Smithsonian Institution, Washington, D.C.
———. Interview by Henry S. Tropp. Tape recording, Washington, D.C., February 18, 1972. No. 196, box 2, files 7, 9, 12, Computer Oral History Collection, Smithsonian Institution, Washington, D.C.
Auerbach, Isaac. Interview by Nancy Stern. Tape recording, April 10, 1978. OH 2, Charles Babbage Institute, University of Minnesota, Minneapolis.

——. Interview by Henry S. Tropp. Tape recording, Washington, D.C., February 17, 1972. No. 196, box 3, file 5, Computer Oral History Collection, Smithsonian Institution, Washington, D.C.

Bartik, Jean, and Frances E. Holberton. Interview by Henry S. Tropp. Tape recording, Washington, D.C., April 27, 1973. No. 196, box 3, file 6, Computer Oral History Collection, Smithsonian Institution, Washington, D.C.

Brainerd, John Grist. Interview. Tape recording. No. 196, January 8, 1970. Box 4, file 6, Computer Oral History Collection, Smithsonian Institution, Washington, D.C.

Burks, Arthur W. Interview by William Aspray. Tape recording, Ann Arbor, Mich., June 20, 1987. OH 136, Charles Babbage Institute, University of Minnesota, Minneapolis.

——. Interview by Christopher Evans for Pioneers of Computing (Science Museum, London). Tape recording, Ann Arbor, Mich., 4976. OH 78, Charles Babbage Institute, University of Minnesota, Minneapolis.

Burks, Arthur W., and Alice R. Burks. Interview by Nancy Stern. Tape recording, Ann Arbor, Mich., June 20, 1980. OH 75, Charles Babbage Institute, University of Minnesota, Minneapolis.

Chambers, Carl. Interview by Nancy Stern. Tape recording, Philadelphia, Pa., November 22, 1977. OH 7, Charles Babbage Institute, University of Minnesota, Minneapolis.

Eckert, J. Presper. Interview by David K. Allison. Tape recording, Washington, D.C., February 2, 1988. National Museum of American History, Smithsonian Institution, Washington, D.C.

——. Interview in *Pennsylvania Triangle.* University of Pennsylvania, Philadelphia, March 1962.

——. Interview by Nancy Stern. Tape recording, Blue Bell, Penn., October 28, 1977. OH 13, Charles Babbage Institute, University of Minnesota, Minneapolis.

Eckert, J. Presper, Kathleen Mauchly Antonelli, William Cleaver, and James McNulty. Interview by Nancy Stern. Tape recording, Philadelphia, Pa., January 23, 1980. OH 11, Charles Babbage Institute, University of Minnesota, Minneapolis.

Elbourn, Robert D. Interview by Richard R. Mertz. Tape recording, March 23, 1971. No. 196, box 6, file 10, Computer Oral History Collection, Smithsonian Institution, Washington, D.C.

Fox, Margaret. Interview by James Ross. Tape recording, Minneapolis,

Minn., April 13, 1983. OH 49, Charles Babbage Institute, University of Minnesota, Minneapolis.

Goldstine, Herman. Interview by Nancy Stern. Tape recording, Princeton, N.J., August 11, 1980. OH 18, Charles Babbage Institute, University of Minnesota, Minneapolis.

Holberton, Frances E. Interview by James Ross. Tape recording, Potomac, Md., April 14, 1983. OH 50, Charles Babbage Institute, University of Minnesota, Minneapolis.

Hopper, Grace. Interview. Tape recording, ca. 1976. OH 81, Charles Babbage Institute, University of Minnesota, Minneapolis.

Masterson, Earl Edgar. Interview by William Aspray and Robbin Clamons. Tape recording, St. Louis Park, Minn., June 30, 1986. OH 115, Charles Babbage Institute, University of Minnesota, Minneapolis.

Mauchly, John W. Interview by Esther Carr. Videotape recording, Ambler, Pa., 1978.

———. Interview by Christopher Evans for Pioneers of Computing (Science Museum, London). Tape recording, Philadelphia, Pa., ca. 1976. OH 26, Charles Babbage Institute, University of Minnesota, Minneapolis.

———. Interview by Henry S. Tropp. Tape recording, February 6, 1973. No. 196, Computer Oral History Collection, Smithsonian Institution, Washington, D.C.

McDonald, Robert Emmett. Interview by James Ross. Tape recording, Minneapolis, Minn., December 16, 1982. OH 45, Charles Babbage Institute, University of Minnesota, Minneapolis.

Metropolis, Nicholas C. Interview by William Aspray. Tape recording, Los Alamos, N.M., May 29, 1987. OH 135, Charles Babbage Institute, University of Minnesota, Minneapolis.

Travis, Irven. Interview by Nancy Stern. Tape recording, Paoli, Penn., October 21, 1977. OH 36, Charles Babbage Institute, University of Minnesota, Minneapolis.

Warren, S. Reid. Interview by Nancy Stern. Tape recording, Philadelphia, Penn., October 5, 1977. OH 38, Charles Babbage Institute, University of Minnesota, Minneapolis.

Interviews by the Author

Antonelli, Kathleen Mauchly. Aberdeen, Md., and Ambler, Pa., November 14, 1996, and April 29, 1997.

Bartik, Jean. Aberdeen, Md., and Ambler, Pa., November 14, 1997, and April 29, 1997.
Butler, Lila. Aberdeen, Md., April 29, 1997.
Chernow, Joseph. Philadelphia, Pa., November 12, 1996.
Chu, Chuan. Telephone, May 10, 1997.
Davis, Jack. Quakertown, Pa., May 1, 1997.
Eckert, Judith. Aberdeen, Md., Gladwyne, Pa., and Ambler, Pa., November 13, 1996, and April 28, 1997.
Gluck, Simon E. Aberdeen, Md., April 30, 1996.
Goldstine, Herman H. Aberdeen, Md., September 13, 1996, and November 13, 1996.
Holberton, Elizabeth Snyder. Aberdeen, Md., November 14, 1996.
Huskey, Harry. Aberdeen, Md., November 14, 1996.
Kelleher, Herbert. Dallas, Tex., May 21, 1998.
Miller, Thomas A. Gladwyne, Pa., April 28, 1997.
Sheppard, Brad. Telephone, April 30, 1997.
Showers, Ralph. Telephone, May 1, 1997.

Correspondence

Atanasoff, John V., to John W. Mauchly, January 23, 1941.
Atanasoff to Charles E. Friley, president, Iowa State College, May 15, 1941.
Atanasoff to Mauchly, May 21, 1941.
Atanasoff to Richard Trexler, attorney at Cox, Moore and Olson, May 21, 1941.
Atanasoff to Mauchly, May 31, 1941.
Atanasoff to Trexler, August 5, 1941.
Atanasoff to Mauchly, October 7, 1941.
Ayres, Quincy C., to Atanasoff, October 16, 1941.
Bradbury, Norris E., director, Los Alamos Project, to Major General G. M. Barnes and Colonel Paul N. Gillon, March 18, 1946.
Cleaver, William E., to Jean Bartik, December 14, 1972.
Gillon, Colonel Paul N., to Moore School of Electrical Engineering, April 10, 1946.
Mauchly to Supreme Instruments Corp., Greenwood, Miss., September 27, 1939.
Mauchly to Atanasoff, January 19, 1941.

Mauchly to Atanasoff, February 24, 1941.
Mauchly to Atanasoff, March 31, 1941.
Mauchly to Atanasoff, May 27, 1941.
Mauchly to Atanasoff, June 7, 1941.
Mauchly to Atanasoff, June 22, 1941.
Mauchly to Sundstrand Co., Massachusetts, June 28, 1941.
Mauchly to Atanasoff, September 30, 1941.
Mauchly and J. Presper Eckert Jr. to S. Reid Warren, November 13, 1945.
Pender, Harold, to Eckert and Mauchly, March 22, 1946.
Warren to Mauchly, April 18, 1946.
Warren, memorandum, April 2, 1947.
Mauchly to Donald E. Knuth, Stanford University, September 2, 1971.

Acknowledgments

This book is the result of the enthusiasm of several people who saw the need for the telling of this story. It never would have come about without their cooperation and assistance.

First and foremost, I am indebted to Nancy Miller for the idea, and to Carol Mann for bringing it to me. Jackie Johnson was a patient soul and thoughtful editor who improved the manuscript. George Gibson was the guiding light.

Jean Bartik was a fountain of information and a burst of energy who spurred me on several times during research and writing. She tracked down ENIAC veterans across the country and pulled together many loose ends. Kathleen Mauchly Antonelli and Judy Eckert both were wonderful sources and devoted keepers of the flame who opened their homes, attics, basements, and memories to me. Herman Goldstine, too, was a patient participant who has an insightful sense of history.

Esther Carr very kindly let me view more than twelve hours of videotape she made of John Mauchly shortly before his death. Her videotaping of Mauchly is a treasure that unfortunately sits in her safe-deposit box. Perhaps this book will help the tapes find a more visible home.

Tom Miller, a close friend of the Eckert family, has become a skilled organizer of Pres Eckert's papers, and he helped assemble pieces of the history.

Kathy Kleiman, who has undertaken a marvelous film project to document the contribution of the women who programmed ENIAC, was a tremendous help to me with research. Her film will undoubtedly be a very important part of women's history, as well as computing history.

Archivists, librarians, and historians at four key institutions offered enormous assistance and guidance. I am indebted to Bruce Bruemmer and Kevin Corbitt at the Charles Babbage Institute in Minneapolis; Paul Shaffer, Mark Frazier Lloyd, and Gail Pietrzyk at the University of Pennsylvania in Philadelphia; Paul Ceruzzi and Allison Oswald at the Smithsonian Institution in Washington, D.C.; and Marjarie McFinch at the Hagley Museum and Library in Wilmington, Delaware.

Friends and colleagues were invaluable sources of information and support, including Tom Petzinger Jr., Rob Tomsho, and Sam Howe Verhovek. Paul Steiger, Dan Hertzberg, and Jim Pensiero at the *Wall Street Journal* graciously gave me the time I needed.

Brad Blumenthal helped jump-start the project and pointed me in the right direction by mapping out a wealth of resources through the Internet. His guidance and computing expertise were an enormous help, and I thank him and Lynn Blumenthal for the careful reading they gave an early draft. Tad Bartimus once again was friend and coach.

As always, I counted on my wife, Karen, for wisdom and editing. And as usual, she had to remind me that not even *Winnie-the-Pooh* begins at the beginning. This was a team effort; her wonderful touch is reflected in many pages of this book.

Index

Index

Index

Index

Index